D1091545

Digital Signal Processing

APPLICATIONS TO
COMMUNICATIONS AND ALGEBRAIC
CODING THEORIES

WITHDRAWN

Digital Signal Processing

APPLICATIONS TO COMMUNICATIONS AND ALGEBRAIC CODING THEORIES

Salvatore D. Morgera

Department of Electrical Engineering
McGill University
Montreal, Quebec
Canada

Hari Krishna

Department of Electrical Engineering
Syracuse University
Syracuse, New York

Tennessee Tech. Library
Cookeville, Tenn.

ACADEMIC PRESS, INC.

Harcourt Brace Jovanovich, Publishers

Boston San Diego New York
Berkeley London Sydney
Tokyo Toronto

389850

Copyright © 1989 by Academic Press, Inc.
All rights reserved.
No part of this publication may be reproduced or
transmitted in any form or by any means, electronic
or mechanical, including photocopy, recording, or
any information storage and retrieval system, without
permission in writing from the publisher.

ACADEMIC PRESS, INC.
1250 Sixth Avenue, San Diego, CA 92101

United Kingdom Edition published by
ACADEMIC PRESS, INC. (LONDON) LTD.
24–28 Oval Road, London NW1 7DX

Library of Congress Cataloging-in-Publication Data

Morgera, Salvatore D. (Salvatore Domenic), Date–
   Digital signal processing.

   Bibliography; p.
   Includes index.
   1. Signal processing–Digital techniques. 2. Coding
theory. I. Krishna, Hari, Date–   . II. Title.
TK5102.5.M66 1988   621.38′043   88–24216
ISBN 0-12-506995-2

Printed in the United States of America
89 90 91 92   9 8 7 6 5 4 3 2 1

TO MY PARENTS
BIANCA AND SALVATORE
AND MY DEAR WIFE
MISCHA
\- SDM

TO MY SISTER
ANURADHA AGGARWAL
\- HK

Contents

Preface

The field of digital signal processing has undergone extraordinary and rapid development over the last 20 years. It is a field, born of the numerical methods and creative genius of such men as Issac Newton (1642-1727) and Carl Friedrich Gauss (1777-1855) and spurred by the advent of high speed digital computing devices, which now finds applications in biomedical data processing; speech processing; radio astronomy; seismic, radar, and sonar signal processing; data communications; and image processing. Digital signal processing is an exciting field of study due to the fact that it is highly applicable and yet rich in its dependence on certain profound mathematics.

About 10 years ago, it became apparent that number theory and, in particular, the theory of congruences introduced by Gauss at age 24 in his book *Disquisitiones Arithmeticae* was destined to have an impact on the field of digital signal processing. The additional names of brilliant mathematicians such as Euclid (300 B.C.); Pierre de Fermat (1601-1665), a lawyer by profession; and Leonard Euler (1707-1783) along with samplings of their exceptional work became much better known to researchers in digital signal processing. A flurry of activity ensued in the development of interesting and computationally efficient digital signal processing algorithms in which sampled data indices or amplitudes were treated in a number-theoretic manner and in which sampled data was transformed according to so-called rectangular linear transformations. Central to many of these techniques is the Chinese Remainder Theorem, a method dating back to antiquity for partitioning a large problem into a number of smaller subproblems. This time of resurgence was important and valuable for digital signal processing, as it strengthened and reestablished the threads back to classical mathematics and opened up numerous applications for applied mathematicians and arithmetic complexity theorists.

Arithmetic complexity theory deals essentially with the study of the limits of the efficiency of algorithms. Keratsuba and Toom, Cooley and Tukey, and Strassen provided the foundations for arithmetic complexity theory during the 1960's with their discovery of algorithms for the multiplication of two integers, computation of the discrete Fourier transformation, and the computation of the product of two matrices, respectively. Deep study of arithmetic complexity theory requires an understanding of linear algebra and the structures of abstract algebra: groups, rings, and fields.

Another area of research requiring a similar background is that of algebraic coding theory, a discipline of information theory. The birth of information theory can be traced back to the late 1940's in Shannon's work dealing with the reliable transmission of information across noisy channels and the important concept of channel capacity. Since that time, much effort has been directed toward the discovery of error-correcting codes having desirable properties and the design of computationally efficient algebraic decoders. Interesting connections between algebraic coding theory and digital signal processing exist; for example, it is known that the decoding of Reed-Solomon and Bose-Chaudhuri-Hocquenghem codes (subfield subcodes of Reed-Solomon codes having the same design minimum distance) using the Berlekamp-Massey algorithm can be viewed as a problem in recursively designing a minimum length linear feedback shift register or autoregressive filter which generates the known sequence of syndrome symbols computed over a finite field using received symbols. Much effort has been directed since the early 1960's toward the design of encoding and decoding schemes which could be easily implemented. The search for codes having reasonably good properties, coupled with an attractiveness in terms of efficient decoding methods is certainly one very practical reason which drew us into the realm of research described in this book. The other reason is solely theoretical and relates simply to the elegance and beauty of algebra, in particular, the theory of finite fields and the opportunity to apply this theory to digital signal processing problems.

The intent of this book is to explore one intersection of the above three disciplines: the design of computationally efficient digital signal processing algorithms over finite fields and the relation of these algorithms to algebraic error-correcting codes. The first goal of the authors was to write a book which felt comfortable in the hands of engineers, mathematicians, and computer scientists carrying out research in their respective areas. We hope that we have achieved this. The second goal of the authors was to make a statement regarding the importance of developing a multidisciplinary outlook on research problems. This is not to say that a broad perspective necessarily insures significant results or is necessarily desirable, since often times it is within a narrow perspective that a coherent pattern of development can be seen. It is to say, however, that at any depth in any problem there are bridges leading in and out. It is sometimes rewarding from both intellectual and practical viewpoints to traverse one or several bridges to new shores. This may require, as it does in this book, some familiarity with abstract algebra for engineers or algebraic coding theory for mathematicians. The third, final, and, perhaps most ambitious goal of the authors was to demonstrate that abstract beginnings can lead to practical endings. This part of the journey required years for us to make, and we are pleased to share it with you. There is still much to do, as in the words of Boris Pasternak (1890-1960), *"What is laid down, ordered, factual, is never enough to embrace the whole truth: life always spills over the rim of every cup."*

SALVATORE DOMENIC MORGERA
Montréal

Acknowledgements

This book would not have been possible without the continued research support of SDM by the Natural Sciences and Engineering Research Council of Canada (NSERC) Grant A0912 and the Québec Fonds pour la Formation de Chercheurs et L'Aide à la Recherche (FCAR) Grant EQ-0350. The initial interest in problems of the type described in this book arose for SDM as a result of taking a course entitled "Complexity of Functions" taught by Charles (Chuck) Fiduccia at Brown University in the Spring Semester of 1971 at the urging of John Savage. Both are due a debt of gratitude for their excellence in research and their willingness to share ideas and thoughts. SDM would also like to acknowledge the assistance of and stimulating association with Meghanad Wagh, a Postdoctoral Fellow at Concordia University, Montréal, during the years 1979-1980, and now with Lehigh University. Many of the ideas found in Chapters 3 and 4 were worked out in joint collaboration with Meghanad Wagh; his competence and perserverance are gratefully appreciated. During the years that SDM was at Concordia University (1978 - 1985), two men were instrumental in their continued encouragement of this work and in the example they set by virtue of their own international calibre research; they are M.N.S. Swamy, Dean of the Faculty of Engineering and Computer Science, and Andreas Antoniou, Chairman of the Department of Electrical Engineering and now with the University of Victoria, British Columbia. SDM will always remain grateful for his association with these scientists. During these years, HK also completed his Ph.D. studies on topics treated in Chapters 5 and 6.

During the years (1985 -) at McGill University, Montréal, and Syracuse University, Syracuse, NY, the final manuscript for this book was completed. SDM would like to acknowledge the computer simulation work found in Chapter 6 and the diligence of Vitalice Oduol. Several discussions with John Labute of the Mathematics and Statistics Department at McGill are also gratefully acknowledged and helped to place Chapter 4 in a proper perspective. HK would like to aknowledge the support and encouragement of his academic activities by Theodore A. Bickart, Dean of the College of Engineering, and Norman Balabanian, Chairman of the Department of Electrical and Computer Engineering.

Many colleagues have provided insight and interest in our results over the years; we are particularly grateful to Shu Lin of the University of Hawaii at Manoa and Daniel Costello of Notre Dame. Other colleagues who have been instrumental in guiding our thoughts are Charles Rader, James McClellan, Bernard Picinbono, and Henri Nussbaumer. From 1987 to the present, SDM expresses gratitude to Bell Canada, Ottawa, and in particular, Don Assaff, Al Delorenzi, Eileen Brown, Lynton Auld, and Steve Rowlandson for their research contract support, stimulating discussions, and opportunity to bring some of our ideas to practice.

The extremely competent work of Anna Cianci and Edith Provost of McGill University is what transformed handwritten and sometimes barely readable pages into the electronic manuscript. The authors would also like to express their appreciation to Alice Peters, Paul Price, Amy Strong, and the staff at Academic Press, Boston, for helping us to make this all happen. The patience, love, and understanding of our families throughout the process of writing and revision is most gratefully acknowledged.

List of Figures

List of Tables

Chapter 1

Overview and Perspective

1.1 Introduction

Digital signal processing is the study of discrete time signals and systems. In this connection, two fundamental signal processing techniques employed for modeling, synthesis, and analysis are *periodic* and *aperiodic convolution*. The former technique is employed to describe the output of a linear-time-invariant finite impulse response (LTI FIR) system to a periodic input; the latter is used for predicting system response to nonperiodic inputs and computing polynomial and large integer products. Moreover, under certain conditions, signal processing transforms and, in particular, the discrete Fourier transform (DFT) can be computed using periodic convolution. We are primarily interested in efficient algorithms for periodic and aperiodic convolution; these algorithms compute quantities which are in the field generated by their inputs.

Let $\mathcal{F}$ be the ground field or field of constants. The functions to be computed are elements of $\mathcal{E} = \mathcal{F}(\mathcal{I})$, the extension of $\mathcal{F}$ by the finite set of indeterminates (inputs) $\mathcal{I}$. We approach the problem of designing efficient algorithms for convolution within the general framework of the system Ψ of bilinear forms given by

$$\psi_t = \sum_{i=1}^{r} \sum_{j=1}^{s} a_{tij}\, x_i y_j, \quad t = 1, 2, \ldots, k; \; a_{tij} \epsilon \mathcal{F}. \qquad (1.1)$$

The indeterminates x_i and y_j are thought of as inputs to an algorithm for computing Ψ, with the constants a_{tij} being independent of the in-

1

puts. The algorithms for periodic and aperiodic convolution that we develop and discuss stem from computations of specific systems of the general type (1.1) and will be of the *bilinear, noncommutative*(NC) variety. Such algorithms take the form $\psi = C(A\mathbf{x} \bullet B\mathbf{y})$, where the matrices A, B, and C have elements from $\mathcal{F}$; the indeterminates x_i and y_j form the vectors $\mathbf{x}$ and $\mathbf{y}$, respectively; and the outputs ψ_t form the vector ψ. The symbol $\bullet$ denotes component-by-component multiplication of the $\mathcal{F}$-linear form vectors $A\mathbf{x}$ and $B\mathbf{y}$. Writing $\psi = C\mathbf{m}$, our interest is principally in reducing the number of *nonscalar* multiplications required by the NC algorithm; this is equivalent to finding a (straight-line) algorithm having the smallest *bilinear length* or length of the vector $\mathbf{m}$. To succeed in this, we bring to bear certain results from arithmetic complexity theory, a field of research rich in its own right. We feel that such a partnership of digital signal processing and computational complexity theory is entirely logical and can assist in reducing the heavy computational burden placed on even the largest computers by present and future signal processing problems. In our work, we have even gone further than this by exploring the intimate manner in which bilinear algorithms for computing (1.1) are related to algebraic coding theory and, thereby, also find practical application in digital communications systems. As the chapters unfold, we hope that the reader will sense the promise of this multidisciplinary undertaking. We use the remainder of this chapter to introduce some notation and concepts and to take the opportunity to touch on various issues either not treated or treated differently in subsequent chapters.

1.2   Breadth and Depth

The aperiodic convolution, written as $\phi = \mathbf{z} \odot \mathbf{y}$, of two indeterminate sequences, $z_0, z_1, \ldots, z_{k-1}$, and $y_0, y_1, \ldots, y_{d-1}$, is given by

$$\phi_t = \sum_{j=0}^{t} z_j y_{t-j}, \quad t = 0, 1, 2, \ldots, k + d - 2, \qquad (1.2)$$

where it is understood that $z_j = 0$, $j > k - 1$, and $y_i = 0$, $i > d - 1$. Let $Z(u) = \sum_{i=0}^{k-1} z_i u^i$ and $Y(u) = \sum_{j=0}^{d-1} y_j u^j$; then, the computation (1.2) may equivalently be written as the polynomial product

$$\Phi(u) = Z(u)Y(u). \qquad (1.3)$$

The polynomial $\Phi(u)$ is $\Phi(u) = \sum_{t=0}^{k+d-2} \phi_t u^t$, where the coefficients of $\Phi(u)$ are the entries of $\phi = \mathbf{z} \odot \mathbf{y}$. We see that the computation of the $N = k + d - 1$ point aperiodic convolution (1.2) is equivalent to the computation of the N coefficients of the polynomial $\Phi(u)$. This computation is a generalization of the approach taken for the multiplication of two large integers, one of the guiding problems in arithmetic complexity theory. A straightforward approach to computing (1.2) or (1.3) requires approximately kd multiplications and additions. This number is larger, however, than the number of *linearly independent* (*l.i.*) coefficients over $\mathcal{F}$ of $\Phi(u)$, a number which is easily shown to be $k+d-1$. Can an algorithm for computing (1.2) be devised which requires only $k+d-1$ multiplications? The answer is yes, if the cardinality of $\mathcal{F}$, $|\mathcal{F}|$, is at least $k + d - 1$. Although we treat this problem in Chapter 3, it is instructive to illustrate here, using a different approach, the nature of a *minimal algorithm* for computing (1.2) or (1.3).

Assume $|\mathcal{F}| \geq k + d - 1$; this allows us to find $k + d - 1$ distinct elements, $a_0, a_1, a_2, \ldots, a_{k+d-2}$, of $\mathcal{F}$. From (1.3), we have

$$\Phi(a_\ell) = Z(a_\ell)Y(a_\ell) = \sum_{t=0}^{k+d-2} \left(\sum_{i+j=t} z_j y_i \right) a_\ell^t, \; \ell = 0, 1, 2, \ldots, k+d-2,$$

$$(1.4)$$

which may be written as

$$\begin{bmatrix} \Phi(a_0) \\ \Phi(a_1) \\ \vdots \\ \Phi(a_{k+d-2}) \end{bmatrix} = \begin{bmatrix} 1 & a_0 & a_0^2 & \cdots & a_0^{k+d-2} \\ 1 & a_1 & a_1^2 & \cdots & a_1^{k+d-2} \\ \vdots & \vdots & \vdots & & \vdots \\ 1 & a_{k+d-2} & a_{k+d-2}^2 & \cdots & a_{k+d-2}^{k+d-2} \end{bmatrix} \begin{bmatrix} \phi_0 \\ \phi_1 \\ \vdots \\ \phi_{k+d-2} \end{bmatrix}.$$

$$(1.5)$$

The polynomial evaluation required in (1.4) could be carried out using *Horner's method*, itself one of the seeds of algebraic complexity theory. The matrix of coefficients $a_0, a_1, a_2, \ldots, a_{k+d-2}$, is a *Vandermonde matrix* and is nonsingular, due to the fact that the selected elements of $\mathcal{F}$ are distinct. Writing (1.5) as $\mathbf{P} = A_N \phi$, we have $\phi = A_N^{-1} \mathbf{P}$. The approach taken is one of *Lagrange interpolation* with interpolation points $a_0, a_1, \ldots, a_{k+d-2}$. Carefully counting the number of multiplications in $\mathcal{F}(I)$, we find that only $k + d - 1$ are required. As is conventional, we do not count so-called *scalar* multiplications by elements of $\mathcal{F}$. If $|\mathcal{F}|$ is

sufficiently large relative to k and d, it is possible to choose the inter-
polation points so that the scalar multiplications are relatively simple.
We note that the number of additions does depend on the selection of
interpolating points. If $|\mathcal{F}|$ is infinite, the number of multiplications
is still $k + d - 1$; this is the theoretical minimum number required to
compute (1.2) or (1.3). If $|\mathcal{F}| \le k+d-2$, the number of multiplications
is indeed greater than $k + d - 1$.

We shall provide computationally efficient algorithms based on the
Chinese Remainder Theorem (CRT) and some novel ideas principally
for the case $|\mathcal{F}| < k + d - 1$. The CRT plays an important role in the
computation of (1.3) if we compute $\Phi(u)$ as $\Phi(u) = Z(u)Y(u) \bmod P(u)$,
where deg $[P(u)] \ge k + d - 1$. If $P(u) = \prod_{i=1}^{L} P_i^{l_i}(u)$, where the
$P_i(u)$ are distinct irreducible polynomials over $\mathcal{F}$ and if the cardinality,
$|\mathcal{F}| < 2 \max \{ \deg [P_i^{l_i}(u)] \} - 2$, then the minimum number of multiplications required to compute (1.3) is $at\ least\ 2 \sum_{i=1}^{L} \deg [P_i^{l_i}(u)] - L$.
The overall algorithm may be written as the $direct\ sum$ of L smaller
algorithms, each associated with a computation modulo $P_i(u)$. We
sometimes write the natural isomorphy of algebras as

$$\mathcal{F}[u]/(P(u)) \cong \mathcal{F}[u]/(P_1^{l_1}(u)) \oplus \cdots \oplus \mathcal{F}[u]/(P_L^{l_L}(u)),$$

where the symbol $\oplus$ is used to denote the direct sum.

Algorithms of the type described above can be employed in at least
two ways. One way of employing these algorithms is for efficiently
performing one of the most frequent signal processing operations, that
of FIR filtering. Consider a specific case of the system Ψ in (1.1), i.e.,

$$\psi = \begin{bmatrix} x_0 & x_1 & \cdots & x_{d-1} \\ x_1 & x_2 & \cdots & x_d \\ \vdots & \vdots & & \vdots \\ x_{k-1} & x_k & \cdots & x_{k+d-2} \end{bmatrix} \begin{bmatrix} y_0 \\ y_1 \\ \vdots \\ y_{d-1} \end{bmatrix}, \tag{1.6}$$

which corresponds to a computation of the first k outputs of an FIR
filter having input sequence $\{x_0, x_1, x_2, \ldots\}$ and impulse response sequence $\{y_0, y_1, \ldots, y_{d-1}\}$. Let the NC algorithm for computing Ψ be
$\psi = C(Ax \bullet By)$, where A is $(n \times N)$-dimensional, B is $(n \times d)$-
dimensional, and C is $(k \times n)$-dimensional, with n being the required
number of multiplications and $N = k + d - 1$. One of the $transposes$
of (1.6) corresponds to the A-$dual$ system Φ of Ψ and to the computation (1.2); this NC algorithm has the form $\phi = A^T (C^T z \bullet By)$. Both

algorithms possess the *same* number of multiplications, n; thus, efficient algorithms for computing aperiodic convolution can be used for FIR filtering applications. The important concept of the duality of a computation will be treated in Chapter 2.

Another way of employing these algorithms is for designing linear error-correcting codes with symbols from $\mathcal{F}$. To briefly illustrate the wide ranging possibilities, we simply note here that if $\phi = A^T(C^T \mathbf{z} \bullet B\mathbf{y})$ is an NC algorithm for computing an aperiodic convolution $\phi = \mathbf{z} \odot \mathbf{y}$, then the *generator matrix* of the linear $(n, k, \bar{d})$ code is given by C. We call $\bar{d}$ the *actual* minimum distance of the code; it is related to the *design* minimum distance d of the code by the inequality $\bar{d} \geq d$. Note that the codeword length, n, is equal to the number of multiplications required for computing an $N = k + d - 1$ point aperiodic convolution. When $n = k + d - 1$, the algorithms developed for aperiodic convolution lead to *maximum-distance-separable* codes and when $n = kd$, to *repetition* codes. Minimizing n for a given k and d is another impetus for the design of efficient aperiodic convolution algorithms. As an example of one result which binds together digital signal processing, arithmetic complexity theory, and algebraic coding theory, we show that the aperiodic convolution of two sequences of lengths k and d can be computed in n multiplications over $\mathcal{F}$ only if there exist $(n, k, \bar{d})$ and $(n, d, \bar{k})$ linear codes over $\mathcal{F}$, where $\bar{d} \geq d$ and $\bar{k} \geq k$. The decoding procedure and a practical application of the linear codes obtained from bilinear NC algorithms are treated at length in Chapters 5 and 6, respectively.

Consider now the indeterminate sequences, $\{x_0, x_1, x_2, \ldots, x_{N-1}\}$ and $\{y_0, y_1, y_2, \ldots, y_{N-1}\}$; the periodic (cyclic) convolution, written as $\psi = \mathbf{x} * \mathbf{y}$, is given by

$$\psi_t = \sum_{j=0}^{N-1} x_{t-j}\, y_j, \quad t = 0, 1, 2, \ldots, N - 1, \tag{1.7}$$

where the subscript $t - j$ is understood to mean $(t - j)$ mod N. The computation (1.7) may equivalently be written as the polynomial product

$$\Psi(u) = X(u)Y(u) \bmod (u^N - 1). \tag{1.8}$$

We see that (1.8) can be computed by first computing (1.3) (with the appropriate change in indeterminate labeling and length) and then reducing the result modulo $u^N - 1$. Efficient algorithms for computing aperiodic (periodic) convolution lead to efficient algorithms for computing periodic (aperiodic) convolution. If $\psi = C(\mathbf{Ax} \bullet \mathbf{By})$ is an NC algorithm for computing (1.7) for two length $N = k + d - 1$ indeterminate sequences, the generator matrix of an $(n, k, \bar{d})$ linear code is obtained by deleting the last $N - k$ consecutive rows of C or the last $N - k$ consecutive columns of A. In the manner previously discussed with regard to the CRT, the multiplicative complexity n of the computation (1.8) can be reduced if $P(u) = u^N - 1$ factors into a number of distinct, irreducible polynomials over $\mathcal{F}$, the factors known as *cyclotomic polynomials*. We now describe a connection between cyclic convolution and the discrete Fourier transform (DFT).

Let $N = p^m - 1$ and $\mathcal{F} = GF(p^m)$, the Galois field of p^m elements, with α a primitive element of $GF(p^m)$. It is well known that the cyclic convolution (1.7) can be obtained from the DFT's X_j and Y_j of x_i and y_i, $i, j = 0, 1, 2, \ldots, N - 1$, respectively, in $GF(p^m)$ as

$$\psi_t = (1/N) \sum_{j=0}^{N-1} X_j Y_j \alpha^{-jt}, \quad t = 0, 1, 2, \ldots, N - 1. \qquad (1.9)$$

In our approach to cyclic convolution design described in Chapter 3, we take a classical Fourier transform-based approach starting from (1.9), in conjunction with a *cyclotomic set* construction. In addition to providing efficient algorithms within which aperiodic convolution can be embedded, these algorithms provide a valuable tool for other applications.

One example of this is the decoding of Reed-Solomon (RS) codes over $GF(2^m)$ using the DFT of length $N = 2^m - 1$. The standard way of implementing the DFT is through the fast Fourier transform (FFT) algorithm. When N factors as $N = \prod_{i=1}^{L} r_i$, the FFT requires $M_1 = N \left(\sum_{i=1}^{L} r_i - L \right)$ multiplications. If r_L factors further as $r_L = s_1 s_2$, then the FFT requires $M_2 = N \left[\sum_{i=1}^{L-1} r_i + s_1 + s_2 - (L+1) \right]$ multiplications. Since $M_1 - M_2 = N \left(r_L - s_1 - s_2 + 1 \right) = N(s_1 s_2 - s_1 - s_2 + 1) > 0$, we have $M_2 < M_1$. We see that the larger the number of integer factors, the more efficient the FFT is; however, in many cases, $N = 2^m - 1$ is not sufficiently composite. In particular, if N is a prime, it is not possible to construct the required DFT algorithm by appropriately combining algorithms of smaller lengths, with the only efficient method being obtained from the relation of a DFT of prime length N to a cyclic convolution of length $N - 1$.

If N is prime, we obtain the DFT X_j of the length N sequence $\{x_0, x_1, x_2, \ldots, x_{N-1}\}$ in $GF(2^m)$ as

$$X_0 = \sum_{i=0}^{N-1} x_i,$$
$$X_{g^{N-1-j}} = x_0 + \psi_j, \quad j = 0, 1, 2, \ldots, N - 2, \tag{1.10}$$

where ψ_j is the jth component of the cyclic convolution of the length $N - 1$ sequences, $\{u_0, u_1, u_2, \ldots, u_{N-2}\}$ and $\{v_0, v_1, v_2, \ldots, v_{N-2}\}$, defined as

$$u_i = x_{g^i},$$
$$v_i = \alpha^{g^{N-1-i}}, \quad i = 0, 1, 2, \ldots, N - 2. \tag{1.11}$$

The quantity g is known as the *generator* of the multiplicative group of integers $\{1, 2, \ldots, N - 1\}$. We note that the computation (1.11) of u_i from x_i involves only a permutation and that the computation (1.10) of X_j from ψ_j involves only additions. The multiplicative complexity arises, therefore, in the evaluation of the cyclic convolution (1.7) of the sequences (1.11). In this case, the cyclic convolution can be performed as a two-dimensional convolution, since $N - 1 = 2^m - 2$, which factors into relatively prime factors 2 and $N' = 2^{m-1} - 1$. Over a field of characteristic two, the cyclic convolution of length 2 requires three multiplications. Chapter 3 provides a design procedure for cyclic convolution algorithms of length N'. As an example, when $N = 31$, a total of 31 multiplications are required for cyclic convolution of length $N' = 15$; therefore, an overall total of 93 multiplications are required for a DFT of length $N = 31$. Guidelines are also provided for reducing the associated number of additions.

Other than mentioning the fundamental role played by the Chinese Remainder Theorem, we have thus far contented ourselves primarily with one-dimensional constructions. Chapter 4 deals with multidimensional constructions for periodic convolution. For notational convenience, let $\mathcal{F}_p = GF(p)$ be the ground field and $\mathcal{F}_{p^e} = GF(p^e)$, a larger field for which e is the least integer such that $N | (p^e - 1)$, where N is the length of the desired cyclic convolution. Let $\mathcal{F}_{p^e}^* = \mathcal{F}_{p^e} - \{0\}$ be cyclically generated by α, a primitive p^eth root of unity. The group algebra over $\mathcal{F}_p$ of the group $\mathcal{F}_{p^e}^*$ may be written as

$$\mathcal{F}_p\left[\mathcal{F}_{p^e}^*\right] \cong \oplus_{i \in S} \; \mathcal{F}_p\,[u]/(P_i^{l_i}\,(u)),$$

where the $P_i(u)$, $i \in S$, are distinct, irreducible factors of $u^{p^{e-1}} - 1$ over $\mathcal{F}_p$. Each element of the direct sum is associated with a computation over a *subfield*. Assume now that N has two mutally prime factors, i.e., $N = N_1 N_2$, where $\gcd(N_1, N_2) = 1$ and e_1 and e_2 are the least integers such that $N_1|(p^{e_1} - 1)$, $N_2|(p^{e_2} - 1)$, with $s = \gcd(e_1, e_2)$. It is now possible to show that the group algebra is a *tensor product*, i.e.,

$$
\begin{aligned}
\mathcal{F}_p\left[\mathcal{F}_{p^e}^*/\mathcal{F}_{p^s}^*\right] &\cong \mathcal{F}_p\left[\mathcal{F}_{p^{e_1}}^*/\mathcal{F}_{p^s}^*\right] \otimes \mathcal{F}_p\left[\mathcal{F}_{p^{e_2}}^*/\mathcal{F}_{p^s}^*\right] \\
&\cong \left[\oplus_{j \in S_1} \mathcal{F}_p[u]/\left(P_j^{l_j}(u)\right)\right] \otimes \left[\oplus_{k \in S_2} \mathcal{F}_p[u]/\left(P_k^{l_k}(u)\right)\right],
\end{aligned}
$$

where the symbol $\otimes$ is used to denote the direct or tensor product. We address the important question of how the individual cyclotomic sets, specifying the roots from $\mathcal{F}_{p^{e_1}}$ and $\mathcal{F}_{p^{e_2}}$ of the distinct irreducible polynomials over $\mathcal{F}_p$, form the roots of the polynomial $u^N - 1$, where $(u^N - 1)|(u^{p^e} - 1)$. It is shown that a *Kronecker product* representation for the cyclic convolution algorithm of length N can only be obtained from cyclic convolution algorithms of lengths N_1 and N_2 if $\gcd(N_1, N_2) = 1$ *and* $\gcd(e_1, e_2) = s = 1$. The form of the algorithm for length N is then

$$\psi = C(A\mathbf{x} \bullet B\mathbf{y}) = (C^{(1)} \otimes C^{(2)})\left[(A^{(2)} \otimes A^{(1)})\mathbf{x} \bullet (B^{(2)} \otimes B^{(1)})\mathbf{y}\right].$$
$$(1.12)$$

The multiplicative complexity for an algorithm of length N is then the *product* of the multiplicative complexities for algorithms of lengths N_1 and N_2. Algorithms formed in this manner are sometimes called *mutally prime factor* algorithms. We prove that for $\gcd(e_1, e_2) = s > 1$, the resulting multiplicative complexity is *less* than the product of complexities and, further, we show how to design such long length algorithms. These results are not only interesting from an algebraic standpoint, but they also provide insight into *product code* constructions and the complexity of cyclic convolution when $\mathcal{F}$ is a finite field and N is arbitrarily large.

We close this chapter by citing a number of general references in each of the three areas touched on in this work. As we do not pretend to be experts in all three areas, we have used these references and expect that most readers will find them helpful. In digital signal processing, we find [1.1] to be a good starting point, providing a practical introduction to number theory and the Chinese Remainder Theorem. For the latter topics, [1.2] is an excellent reference and for algebra, the classical reference [1.3] is most readable. With respect to error-correcting codes, we recommend [1.4] and [1.5], both of which have proved most helpful during the course of this work. Our most often used references in arithmetic complexity theory are [1.6], for a practical view of problems, and [1.7], for a rather nice treatment of more complex topics, tensors, and direct sum and product computations. Additional background material on digital communications systems with retransmission strategies of the type discussed in Chapter 6 may be found in [1.8].

1.3 References

[**1.1**] J. H. McClellan and C. M. Rader, *Number Theory in Digital Signal Processing*, Prentice-Hall, 1979.

[**1.2**] J. Nagell, *Introduction to Number Theory*, Chelsea, 1964.

[**1.3**] B. L. Van der Waerden, *Algebra*, Frederick Ungar Publishing Co., 1970 (7th ed., 2 vols.).

[**1.4**] E. R. Berlekamp, *Algebraic Coding Theory*, McGraw-Hill, 1968.

[**1.5**] I. F. Blake and R. C. Mullin, *The Mathematical Theory of Coding*, Academic Press, 1975.

[**1.6**] S. Winograd, *Arithmetic Complexity of Computations*, SIAM CBMS-NSF regional conference series in applied mathematics, 1980.

[**1.7**] H. F. de Groote, *Lectures on the Complexity of Bilinear Problems*, Springer-Verlag lecture notes in computer science (G. Goos and J. Hartmanis, eds.), 1987.

[**1.8**] S. Lin and D. J. Costello, Jr., *Error Control Coding: Fundamentals and Applications*, Prentice-Hall, 1983.

Chapter 2

Systems of Bilinear Forms

The examples of the previous chapter demonstrate that study of some of the most widely used digital signal processing techniques is synonymous with study of systems of bilinear forms. The intent of this chapter is to briefly review the computational complexity associated with systems of bilinear forms over an extension of a field of constants and to describe a most interesting connection of systems of bilinear forms to error-correcting codes. This chapter is intentionally short so that the reader may fully appreciate the closeness of aspects of digital signal processing and algebraic coding theory when approached from a computational complexity standpoint.

2.1 The Program Model

In the area of arithmetic complexity, algebraic problems such as function evaluation are analyzed in order to ascertain the number of arithmetic operations required by a chosen algorithm. Let $\mathcal{F}$ be the ground field or field of constants. The functions to be computed are elements of $\mathcal{E} = \mathcal{F}(\mathcal{I})$, the extension of $\mathcal{F}$ by multinomial expressions of the elements of the finite set of indeterminates $\mathcal{I}$. In this chapter, the variable names x, y, z will generally be used for elements contained in $\mathcal{I}$. The model of computation we adopt is the *straight-line program model*, for which a computation consists of a sequence of instructions of the type

$$f_i \leftarrow g_i \circ h_i. \tag{2.1}$$

The symbol $\circ$ stands for addition $(+)$, subtraction $(-)$, or multipli-

11

cation ($\times$). The quantity f_i is a variable not appearing in any previous step, and g_i and h_i are either indeterminates, elements of $\mathcal{F}$, or variable names appearing on the left of the arrow at some previous step. An element of $\mathcal{F}$ appearing in a computation is called a constant. A computation computes E, a set of expressions in $\mathcal{E}$, if, for each expression $e \epsilon E$, there is some variable f in the computation such that $f = e$. The interested reader may refer to [2.1] for additional details on this form of program model. We only mention that our interest is with algorithms having *low* complexity (and even then, many times only low *multiplicative* complexity); consequently, we adopt a straight-line program model, since, in general, branching on some arithmetic predicate does not reduce the amount of computation an algorithm performs.

The *multiplicative complexity* of an expression e is defined as the number of instructions of the specific type $f_i \leftarrow g_i \times h_i$ required to compute e. Instructions involving multiplications by elements of $\mathcal{F}$ are not included in the multiplicative complexity count. This is a conventional approach, taken for mathematical convenience (we note that this approach is easily justified if $a \epsilon \mathcal{F}$ is an integer, for which the multiplication by a can be replaced by additions, or even if $a \epsilon \mathcal{F}$ is rational). We also do not include division as an operation, since it is well known that without division, computations of systems of bilinear forms can be reduced to computing linear combinations of products of pairs of linear forms.

2.2 Multiplicative Complexity

A system Ψ of k bilinear forms may be written as

$$\psi_t = \sum_{i=1}^{r} \sum_{j=1}^{s} a_{tij} \, x_i \, y_j, \quad 1 \le t \le k, \ a_{tij} \, \epsilon \, \mathcal{F}. \tag{2.2}$$

The system (2.2) may be equally well represented in the matrix form

$$\psi = X \, \mathbf{y}, \tag{2.3}$$

where $\mathbf{y} = [y_1 \ y_2 \ \cdots \ y_s]^T$ represents a vector of indeterminates and the elements of the $(k \times s)$-dimensional matrix X are $\mathcal{F}$-linear forms in the indeterminates $\{x_1, \ x_2 \ \ldots, \ x_r\}$. We also define $\mathbf{x} = [x_1 \ x_2 \ \cdots \ x_r]^T$.

Now, a *bilinear algorithm* of multiplicative complexity n for computing Ψ over $\mathcal{F}$ is an expression of the type

$$\psi = C(A\mathbf{x} \bullet B\mathbf{y}) = C\mathbf{m}, \qquad (2.4)$$

where the $(n \times r)$, $(n \times s)$, and $(k \times n)$-dimensional matrices A, B, and C, respectively, have entries from $\mathcal{F}$ and the symbol $\bullet$ stands for component-by-component multiplication of vectors [2.2]. In the alternate, but equivalent, expression for ψ, $\mathbf{m} = [m_1\ m_2\ \cdots\ m_n]^T$, with $m_i = \alpha_i \beta_i$, $i = 1, 2, \ldots, n$. The α_i and β_i are $\mathcal{F}$-linear forms in the indeterminates $\{x_1, x_2, \ldots, x_r\}$ and $\{y_1, y_2, \ldots y_s\}$; specifically, we may write, $\alpha_i = \mathbf{A}_i^T \mathbf{x}$, $\beta_i = \mathbf{B}_i^T \mathbf{y}$, $i = 1, 2, \ldots, n$, where $\mathbf{A}_i^T \epsilon \mathcal{F}^r$ and $\mathbf{B}_i^T \epsilon \mathcal{F}^s$ are the ith row of A and B, respectively. In the bilinear algorithm of (2.4), the multiplicative complexity stems directly from the multiplications needed to compute $A\mathbf{x} \bullet B\mathbf{y}$, i.e., the elements m_i, $i = 1, 2, \ldots, n$, of $\mathbf{m}$. The algorithm (2.4) for computing Ψ is said to be *noncommutative* (*NC*). The validity of such *NC* algorithms does not depend on the commutative law of multiplication, thereby offering a wide range of practical usefulness.

It is generally not difficult to devise *NC* algorithms for computing Ψ for which the multiplicative complexity n is much greater than the dimensions of the linear spaces spanned by the rows and columns (over $\mathcal{F}$) of the matrix X in (2.3). On the other hand, the $2n$ $\mathcal{F}$-linear forms, $\{\alpha_1, \alpha_2, \ldots, \alpha_n\}$ and $\{\beta_1, \beta_2, \ldots, \beta_n\}$, will not exist if n is less than the *theoretical minimum* number of multiplications needed to compute Ψ. Let μ denote the theoretical minimum number of multiplications required to compute Ψ. We now state a definition and several results which establish a lower bound for the multiplicative complexity of algorithms used to compute Ψ. The proofs are straightforward; details may be found in [2.2, 2.3].

DEFINITION 1. Let $\mathcal{F}$ be the ground field, $\mathcal{E} = \mathcal{F}(I)$ the extension of $\mathcal{F}$ by the set of indeterminates I, $\mathcal{F}^m$ the m-dimensional vector space over $\mathcal{F}$, and $\mathcal{F}^m(I)$ the m-dimensional vector space over $\mathcal{F}(I)$. A set of vectors $\{\mathbf{v}_1, \mathbf{v}_2, \ldots, \mathbf{v}_n\}$ from $\mathcal{F}^m(I)$ is *linearly independent* modulo $\mathcal{F}^m$, if for $a_i \epsilon \mathcal{F}$, $i = 1, 2, \ldots, n$, $\sum_{i=1}^n a_i \mathbf{v}_i$ in $\mathcal{F}^m$ implies that all the a_i are zero. The *row rank*, $\rho_r(X)$, of a $(k \times s)$-dimensional matrix X modulo $\mathcal{F}^s$ having entries from $\mathcal{F}(I)$ is the number of linearly independent rows of X modulo $\mathcal{F}^s$. The *column rank*, $\rho_c(X)$, of X modulo $\mathcal{F}^k$ is the number of linearly independent columns of X modulo $\mathcal{F}^k$.

THEOREM 2.1 *(A row-oriented lower bound on the number of multiplications). Let $X\mathbf{y}$ be a system of bilinear forms over $\mathcal{F}$, where X is $(k \times s)$-dimensional. If the row rank of X, $\rho_r(X) = \alpha \leq k$, then any straight-line computation of $X\mathbf{y}$ requires at least $\mu = \alpha$ multiplications.*

THEOREM 2.2 *(A column-oriented lower bound on the number of multiplications). Let $X\mathbf{y}$ be a system of bilinear forms over $\mathcal{F}$, where X is $(k \times s)$-dimensional. If the column rank of X, $\rho_c(X) = \beta \leq s$, then any straight-line computation of $X\mathbf{y}$ requires at least $\mu = \beta$ multiplications.*

THEOREM 2.3 *(A row- and column-oriented lower bound on the number of multiplications). Let $X\mathbf{y}$ be a system of bilinear forms over $\mathcal{F}$. If X has a submatrix W with α rows and β columns such that for any vectors $\mathbf{u}\epsilon\mathcal{F}^\alpha$ and $\mathbf{v}\epsilon\mathcal{F}^\beta$, $\mathbf{u}^TW\mathbf{v}$ is an element of $\mathcal{F}$ iff (if and only if) either $\mathbf{u} = \mathbf{0}$ or $\mathbf{v} = \mathbf{0}$, then any straight-line computation of $X\mathbf{y}$ requires at least $\mu = \alpha + \beta - 1$ multiplications.*

From the above, we conclude that $n \geq \mu$ in order to compute Ψ using an NC algorithm (2.4). Within this class of algorithms, the computational superiority of algorithms may be judged (at least from a multiplicative complexity standpoint) by how close n is to μ. Of particular interest in the sequel will be the case for which the row rank $\rho_r(X) = k$; this results in linear independence of the ψ_i, $i = 1, 2, \ldots, k$. By Theorem 2.1, $n \geq k$ and the rank of the matrix C is k. Before leaving this section, we present a simple example illustrating some of the concepts discussed. This example will also provide a means for introducing the notion of a *minimal algorithm*.

EXAMPLE 1.   Let $\mathcal{F} = \mathcal{R}$, the field of real numbers, and $\mathcal{E} = \mathcal{R}(x_1, x_2; y_1, y_2)$. Consider the following system of bilinear forms corresponding to the real and imaginary parts of a complex multiplication $(x_1 + ix_2)(y_1 + iy_2)$:

$$\psi = \begin{bmatrix} x_1 & -x_2 \\ x_2 & x_1 \end{bmatrix} \begin{bmatrix} y_1 \\ y_2 \end{bmatrix}.$$

Using an NC algorithm of the form (2.4), it is not difficult to show that

$$A = \begin{bmatrix} 1 & 0 \\ 1 & 1 \\ 1 & -1 \end{bmatrix}, \quad B = \begin{bmatrix} 1 & 1 \\ 0 & 1 \\ 1 & 0 \end{bmatrix}, \quad C = \begin{bmatrix} 1 & -1 & 0 \\ 1 & 0 & -1 \end{bmatrix},$$

with $\mathbf{m}$ having elements corresponding to multiplications of the form

$$\begin{aligned}
m_1 &= \alpha_1 \beta_1 = x_1 (y_1 + y_2), \\
m_2 &= \alpha_2 \beta_2 = (x_1 + x_2) y_2, \\
m_3 &= \alpha_3 \beta_3 = (x_1 - x_2) y_1.
\end{aligned}$$

The algorithm has $n = 3$ real multiplications (and 5 real additions), and from Theorem 2.3, we find that the minimum multiplicative complexity over $\mathcal{R}$ is $\mu = 3$. Thus, the above bilinear algorithm has minimum multiplicative complexity. As is often the case, the multiplicative complexity depends on the choice of $\mathcal{F}$.

We now define an $(n \times n)$-dimensional diagonal matrix $D_a(\mathbf{x})$ having the $\mathcal{F}$-linear forms, α_i, $i = 1, 2, \ldots, n$, as diagonal elements. The NC algorithm (2.4) may be written as

$$\psi = C\mathbf{m} = C D_a(\mathbf{x}) B\mathbf{y}, \tag{2.5}$$

which, from (2.3) and the fact that (2.5) must hold for any $\mathbf{y}$, implies the decomposition [2.4]

$$X = C \, D_a(\mathbf{x}) B. \tag{2.6}$$

The converse also holds; thus, there is a one-to-one correspondence between NC algorithms $C\mathbf{m}$ for computing ψ and the decomposition of X as in (2.6). If n is the least integer for which the decomposition (2.6) exists, the NC algorithm $C\mathbf{m}$ for computing ψ is said to be minimal. As mentioned earlier, NC algorithms $C\mathbf{m}$ for computing ψ exist, of course, for dimensions of $D_a(\mathbf{x})$ larger than n; such algorithms are nonminimal, but many times easier to formulate. We leave it to the reader to verify the forms of (2.5) and (2.6), with respect to Example 1 for which

$$D_a(\mathbf{x}) = \begin{bmatrix} x_1 & 0 & 0 \\ 0 & x_1 + x_2 & 0 \\ 0 & 0 & x_1 - x_2 \end{bmatrix}.$$

2.3   Dual of a Bilinear Form

Thus far, we have presented an NC algorithm structure for computing Ψ but have made no attempt to offer a strategy for actual algorithm design. We would expect that as n approaches μ, the difficulty of algorithm design would increase. The concept of duality can prove extremely important in this regard as it provides a number of different algorithmic transpositions with the same NC structure, each having the *same* multiplicative complexity as the original computational problem. The notion of the duality of a computation can be developed from the results of [2.5] or [2.6]; a more recent work dealing exclusively with duality is [2.4]. Before we proceed to the concept of duality, we further examine the system Ψ of (2.2). It is also possible to write (2.2) in the form

$$\psi_t = \mathbf{x}^T G_t \mathbf{y}, \tag{2.7}$$

where, for a multiplicative complexity $n \geq \mu$,

$$G_t = \sum_{j=1}^{n} c_{tj} \, \mathbf{A}_j \mathbf{B}_j^T, \qquad 1 \leq t \leq k. \tag{2.8}$$

The matrix G_t is $(r \times s)$-dimensional with entries from $\mathcal{F}$ and $[C]_{tj} = c_{tj} \epsilon \mathcal{F}$, $\mathbf{A}_j \epsilon \mathcal{F}^r$, and $\mathbf{B}_j \epsilon \mathcal{F}^s$ were previously defined. We see from (2.8) that μ is, therefore, the smallest number of rank one matrices $\mathbf{A}_j \mathbf{B}_j^T$ necessary to include G_t in their span. It seems that this observation was first made in [2.7].

Let $\mathbf{z} = [z_1 \ z_2 \ \cdots \ z_k]^T$ be another set of indeterminates from $\mathcal{I}$ which are independent of the indeterminates $\{x_1, \ x_2, \ \cdots, \ x_r\}$ and $\{y_1, \ y_2, \ \cdots, \ y_s\}$. Consider now the *trilinear* form T, often called the *defining function*,

$$T = \mathbf{z}^T X \mathbf{y} = \sum_{t=1}^{k} z_t \psi_t = \sum_{t=1}^{k} \sum_{i=1}^{r} \sum_{j=1}^{s} a_{tij} z_t x_i y_j, \tag{2.9}$$

which, using (2.7) and (2.8) may be written as

$$\mathbf{z}^T X \mathbf{y} = \mathbf{x}^T Z \mathbf{y}, \tag{2.10}$$

where the $(r \times s)$-dimensional matrix Z is given by

$$Z = \sum_{t=1}^{k} \sum_{j=1}^{n} z_t c_{tj} \mathbf{A}_j \mathbf{B}_j^T. \tag{2.11}$$

The matrix Z is seen to have entries which are $\mathcal{F}$-linear forms in the indeterminates $\{z_1, z_2, \cdots, z_k\}$.

We note that the trilinear form is completely symmetric with respect to the three sets of indeterminates $\mathbf{x}, \mathbf{y}$, and $\mathbf{z}$. For example, the above problem is equivalent to the evaluation of the r bilinear forms ϕ_i, $i = 1, 2, \ldots, r$, associated with the $(k \times s)$-dimensional matrices $G_i = (a_{tij})$, $i = 1, 2, \ldots, r$. Call this system Φ, where

$$\phi = Z\mathbf{y}. \tag{2.12}$$

The system Φ is a *dual system* of Ψ and satisfies (2.10). A consequence of duality, discussed in [2.4], is that the multiplicative complexity of a $(k \times l)$ by $(l \times r)$ matrix multiplication is the same as that of a $(k \times r)$ by $(r \times l)$ matrix multiplication. From (2.11) it immediately follows that

$$Z = A^T D_c(\mathbf{z}) B, \tag{2.13}$$

where the $(n \times n)$-dimensional diagonal matrix $D_c(\mathbf{z})$ has $\mathbf{z}^T \mathbf{C}_j$, $j = 1, 2, \ldots, n$, as diagonal elements, where $\mathbf{C}_j \epsilon \mathcal{F}^k$ is the jth column of C. The decomposition (2.13) for the dual system (2.12) should be compared with the decomposition (2.6) for the original system (2.3). With respect to Example 1, we once again leave it to the reader to verify that

$$D_c(\mathbf{z}) = \begin{bmatrix} z_1 + z_2 & 0 & 0 \\ 0 & -z_1 & 0 \\ 0 & 0 & -z_2 \end{bmatrix}$$

for the dual system Φ.

An *NC* algorithm which is the *A-dual* (sometimes called the *P-dual*) of the computation (2.4) can be associated with Φ and from (2.12) and (2.13) has the form $A^T(C^T\mathbf{z} \bullet B\mathbf{y})$. As $C(A\mathbf{x} \bullet B\mathbf{y})$ is a computation of $X\mathbf{y}$, the *A-dual* $A^T(C^T\mathbf{z} \bullet B\mathbf{y})$ is a computation of $Z\mathbf{y}$. Manipulations similar to those above result in another dual of the computation (2.4); this is the *B-dual* (sometimes called the *R-dual*) with the form $B^T(A\mathbf{x} \bullet C^T\mathbf{z})$. The following result from [2.4] will be employed in the next section.

THEOREM 2.4 *There is a computation for the system of expressions represented by $C(A\mathbf{x} \bullet B\mathbf{y})$ having $n \geq \mu$ multiplications iff there is a computation having n multiplications for its A-dual $A^T(C^T\mathbf{z} \bullet B\mathbf{y})$, its B-dual $B^T(A\mathbf{x} \bullet C^T\mathbf{z})$, the indeterminate commuted system $C(B\mathbf{y} \bullet A\mathbf{x})$, and the indeterminate commuted A- and B- dual systems $A^T(B\mathbf{y} \bullet C^T\mathbf{z})$ and $B^T(C^T\mathbf{z} \bullet A\mathbf{x})$, respectively.*

2.4   A Particular Bilinear Form

We now consider a connection between systems of bilinear forms and algebraic coding theory by examining a specific bilinear form. Recall the remark made at the end of Section 2.2 regarding the situation when the row rank $\rho_r(X) = k$. In this case, $n \geq k$ and the rank of C is k; thus, C can be treated as the *generator matrix* of a linear (n, k) code with symbols from $\mathcal{F}$. A typical codeword $\mathbf{c}^T$ is $\mathbf{a}^TC$, where $\mathbf{a}\epsilon\mathcal{F}^k$ consists of information symbols. Using (2.3) and (2.4) we have the bilinear form

$$\mathbf{a}^TX\mathbf{y} = \mathbf{a}^TC\ (A\mathbf{x} \bullet B\mathbf{y}) = \mathbf{a}^TC\mathbf{m} = \mathbf{c}^T\mathbf{m}, \qquad (2.14)$$

from which it is apparent that the *Hamming weight* (number of nonzero components) of $\mathbf{c}$ cannot be less than the multiplicative complexity associated with (2.14), which, by Theorem 2.2 cannot be less than the number of independent columns in $\mathbf{a}^TX$. The following theorem from [2.8] is important in that it connects the multiplicative complexity of *NC* algorithms for computing Ψ to the algebraic code property of *Hamming distance* and, thus, to error-correcting power. It results as a direct consequence of Theorem 2.2 applied to (2.14).

THEOREM 2.5 *Given a system $\psi = Xy$ of k linearly independent bilinear forms, for every computation of the type $\psi = C(Ax \bullet By)$, the $(k \times n)$-dimensional matrix C generates a linear $(n, k, \bar{d})$ code over $\mathcal{F}$ where*

$$\bar{d} \geq d \;\; = \;\; \min_{\substack{a \in \mathcal{F}^k \\ a \neq 0}} \{ \rho_c(a^T X) \}$$

and $\rho_c(a^T X) \leq s$ denotes the number of linearly independent columns in $a^T X$. The quantities $d, \bar{d}$ are the respective design minimum and actual minimum distances of the linear code constructed.

A linear code constructed in the above manner has a codeword length n equal to the multiplicative complexity of the NC algorithm for computing Ψ and an actual minimum distance $\bar{d}$ bounded from below by the minimum number of multiplications required to compute any linear combination of the k forms in Ψ. The general connection between algorithms for computing bilinear forms and linear *binary* codes ($\mathcal{F} = GF(2)$) was first observed we believe in [2.9] and then expanded upon in [2.8]. For given code parameters k and d, we may consider a good code to be one for which n is as small as possible; thus, efficient computational algorithms for Ψ (or the dual Φ) can lead to good codes. Other code properties, such as weight distribution and encoding / decoding procedure, depend on the actual *structure* of the computational algorithm. With these ideas in mind, we turn to a particular bilinear form.

Consider the following system Ψ of bilinear forms for which X is $(k \times d)$-dimensional:

$$\psi = Xy = \begin{bmatrix} x_0 & x_1 & x_2 & \cdots & x_{d-2} & x_{d-1} \\ x_1 & x_2 & x_3 & \cdots & x_{d-1} & x_d \\ x_2 & x_3 & x_4 & \cdots & \cdot & \cdot \\ \vdots & \vdots & \vdots & & \vdots & \vdots \\ x_{k-1} & x_k & x_{k+1} & \cdots & x_{k+d-3} & x_{k+d-2} \end{bmatrix} \begin{bmatrix} y_0 \\ y_1 \\ \vdots \\ y_{d-1} \end{bmatrix}.$$

$$(2.15)$$

We have seen a bilinear form of this type before in Chapter 1 in regard to FIR filtering. In that context, we can think of $\{x_0, \; x_1, \; x_2, \ldots \}$

as a sequence obtained by sampling an input signal at equally spaced
time instants and $\{y_0,\ y_1,\ \ldots,\ y_{d-1}\}$ as the impulse response of the
FIR filter. The system Ψ of (2.15) then computes the first k outputs
$\psi_t = \sum_{j=0}^{d-1} x_{t+j}\, y_j,\ t = 0, 1, \ldots,\ k-1$, of an FIR filter. Perhaps
surprisingly, this system of bilinear forms will now also prove to be
relevant in a quite different context.

It is not difficult to show that for $\forall \mathbf{a} \epsilon \mathcal{F}^k$, $\mathbf{a} \neq \mathbf{0}$, and $\forall \mathbf{b} \epsilon \mathcal{F}^d$, $\mathbf{b} \neq \mathbf{0}$,
we have $\mathbf{a}^T X \mathbf{b} \neq 0$; therefore, the number of linearly independent
columns, $\rho_c(\mathbf{a}^T X) = d$ for all nonzero $\mathbf{a} \epsilon \mathcal{F}^k$. From Theorem 2.5, we
have that if $C(A\mathbf{x} \bullet B\mathbf{y})$ is a computation of Ψ having n multiplications,
then C generates an $(n, k, \bar{d})$ linear code over $\mathcal{F}$ with $\bar{d} \geq d$. Furthermore, by Theorem 2.3, we have $n \geq k + d - 1$ (note $X = W$). If the
actual multiplicative complexity $n = k + d - 1$, then the corresponding linear codes are referred to as *maximum-distance-separable* (MDS)
codes [2.10]. These codes are sometimes referred to as *optimal*, since
an MDS code has the maximum possible distance between codewords
and has a systematic encoder.

There are now (at least) two ways to proceed in developing an NC
algorithm to compute (2.15). We consider

1. Viewing the system Ψ as the *periodic (cyclic)* convolution of two
   length $N = k + d - 1$ sequences and

2. Viewing the A-dual system Φ as an *aperiodic (noncyclic)* convolution of length $N = k + d - 1$.

From the standpoint of digital signal processing (for which we should
also be interested in *additive* complexity), both approaches provide
an impetus for development of equally important efficient periodic and
aperiodic convolution algorithms over the field $\mathcal{F}$. From the standpoint
of algebraic coding theory, extensive design work seems to indicate that
the aperiodic approach (2) leads to linear codes having more desirable
properties, although approach (1) leads to very interesting multidimensional codes and constructions.

With regard to approach (1) above, the cyclic convolution of two
sequences $\mathbf{x} = [x_0 x_1 \cdots x_{N-1}]^T$ and $\mathbf{y} = [y_0 y_1 \cdots y_{N-1}]^T$ of length N is
typically defined by

$$\psi_t' = \sum_{j=0}^{N-1} x_{t-j} y_j, \quad t = 0, 1, \ldots, N-1, \tag{2.16}$$

where the subscript $t - j$ is understood to mean $(t - j)$ mod N. The notation $\psi' = \mathbf{x} * \mathbf{y}$ is used to represent the operations carried out in (2.16). The cyclic *correlation* (also sometimes called a cyclic convolution) is defined as

$$\psi_t'' = \sum_{j=0}^{N-1} x_{t+j} y_j, \quad t = 0, 1, \ldots, N-1, \tag{2.17}$$

where the subscript $t + j$ is understood to mean $(t + j)$ mod N. If we let $y_{N-j} = v_j$, $j = 0, 1, \ldots, N - 1$, and take $y_N = y_0$, we have $\sum_{j=0}^{N-1} x_{t-j} y_j = \sum_{j=0}^{N-1} x_{t+j} v_j$. Let $\mathbf{v} = [v_0 v_1 \cdots v_{N-1}]^T$. Consider *embedding* the system Ψ of (2.15) in a larger N-dimensional system Ψ'' obtained by using (2.17) and appending $k - 1$ zeroes to the indeterminates $\{y_0, y_1, \ldots, y_{d-1}\}$, to obtain

$$\psi'' = \begin{bmatrix} x_0 & x_1 & x_2 & \cdots & x_{k+d-3} & x_{k+d-2} \\ x_1 & x_2 & x_3 & \cdots & x_{k+d-2} & x_0 \\ \vdots & \vdots & \vdots & \ddots & \vdots & \vdots \\ x_{k+d-2} & x_0 & x_1 & \cdots & x_{k+d-4} & x_{k+d-3} \end{bmatrix} \begin{bmatrix} y_0 \\ y_1 \\ \vdots \\ y_{d-1} \\ 0 \\ \vdots \\ 0 \end{bmatrix}. \tag{2.18}$$

We have $N = k + d - 1$ and $\{\psi_t''\} \supseteq \{\psi_t\}$; specifically, $\psi_t = \psi_t''$, $t = 0, 1, \ldots, k-1$. The system (2.18) represents the manner in which Ψ can be computed as a cyclic correlation. From our previous discussion, we also see that Ψ may be computed as a cyclic convolution, since $\{\psi_t'\} \supseteq \{\psi_t\}$, where $\psi' = \mathbf{x} * \mathbf{v}$ and corresponds to the system of bilinear forms

$$\psi' = \begin{bmatrix} x_0 & x_{k+d-2} & x_{k+d-3} & \cdots & x_2 & x_1 \\ x_1 & x_0 & x_{k+d-2} & \cdots & x_3 & x_2 \\ \vdots & \vdots & \vdots & \ddots & \vdots & \vdots \\ x_{k+d-2} & x_{k+d-3} & x_{k+d-4} & \cdots & x_1 & x_0 \end{bmatrix} \begin{bmatrix} y_0 \\ 0 \\ \vdots \\ 0 \\ y_{d-1} \\ \vdots \\ y_2 \\ y_1 \end{bmatrix}. \tag{2.19}$$

We note that both indeterminate coefficient matrices appearing in
(2.18) and (2.19) are of *circulant*, or cyclic, form. In the first part
of Chapter 3, efficient NC algorithms of the type $C(\mathbf{Ax} \bullet \mathbf{By})$ for
computing length N cyclic convolutions over finite fields are derived
and discussed.

Now, with regard to approach (2) above, let us examine Φ, the
A-dual system of Ψ. The trilinear form T is given by

$$T = \mathbf{z}^T X \mathbf{y} = \sum_{t=0}^{k-1} z_t \sum_{j=0}^{d-1} x_{t+j} y_j = \sum_{t=0}^{k+d-2} x_t \sum_{j=0}^{t} z_j y_{t-j}, \qquad (2.20)$$

where it is understood that $z_t = 0$ for $t > k-1$ and $y_j = 0$ for $j > d-1$.
Using the method described earlier for extracting Φ from the trilinear
form, we find that

$$\phi_t = \sum_{j=0}^{t} z_j y_{t-j}, \quad t = 0, 1, \ldots, N-1, \qquad (2.21)$$

where $N = k + d - 1$. The bilinear forms associated with the A-dual
are, therefore, computable as the elements of the aperiodic convolution
$\phi = \mathbf{z} \odot \mathbf{y}$ of the length k sequence $\mathbf{z}$ and the length d sequence $\mathbf{y}$. The
aperiodic convolution may also be written in matrix form using the
$(N \times d)$-dimensional matrix Z as

$$\phi = Z\mathbf{y} = \begin{bmatrix} z_0 & 0 & 0 & \cdots & 0 \\ z_1 & z_0 & 0 & & \vdots \\ z_2 & z_1 & z_0 & & 0 \\ \vdots & z_2 & z_1 & \vdots & \vdots \\ z_{k-1} & \vdots & z_2 & & z_0 \\ \vdots & z_{k-1} & & & z_1 \\ 0 & \vdots & \vdots & & \vdots \\ 0 & 0 & 0 & \cdots & z_{k-1} \end{bmatrix} \begin{bmatrix} y_0 \\ y_1 \\ \vdots \\ y_{d-1} \end{bmatrix}. \qquad (2.22)$$

Define the *generating polynomials* (one-sided Z-transform) of the
two indeterminate sequences as

$$Z(u) = \sum_{t=0}^{k-1} z_t u^t \text{ and } Y(u) = \sum_{t=0}^{d-1} y_t u^t. \tag{2.23}$$

The generating polynomial of the sequence ϕ_t, $t = 0, 1, \ldots, N-1$, is then given by

$$\Phi(u) = \sum_{t=0}^{N-1} \phi_t u^t = Z(u)\, Y(u), \tag{2.24}$$

which permits us to represent the A-dual Φ as a polynomial product. If the A-dual is computed by the NC algorithm $A^T(C^T \mathbf{z} \bullet B\mathbf{y})$, the generator matrix of the linear $(n, k, \bar{d})$ code is still C, although its placement in the computation vis-à-vis $C(A\mathbf{x} \bullet B\mathbf{y})$ has changed.

We also observe that the system Φ of (2.22) remains unchanged if the $(N \times d)$-dimensional matrix Z is appropriately modified to an $(N \times N)$-dimensional matrix as follows:

$$\phi = \begin{bmatrix} z_0 & 0 & 0 & \cdots & 0 & z_{k-1} & \cdots & z_1 \\ z_1 & z_0 & 0 & & \vdots & 0 & & z_2 \\ z_2 & z_1 & z_0 & & 0 & & & \vdots \\ \vdots & z_2 & z_1 & \vdots & \vdots & \vdots & & z_{k-1} \\ z_{k-1} & \vdots & z_2 & & z_0 & & & 0 \\ \vdots & z_{k-1} & & & z_1 & & & \vdots \\ 0 & \vdots & \vdots & & \vdots & & z_0 & 0 \\ 0 & 0 & 0 & \cdots & z_{k-1} & \cdots & \cdots & z_0 \end{bmatrix} \begin{bmatrix} y_0 \\ y_1 \\ \vdots \\ y_{d-1} \\ 0 \\ \vdots \\ 0 \\ 0 \end{bmatrix}. \tag{2.25}$$

The above system of bilinear forms demonstrates that the aperiodic convolution of a length k and a length d sequence may also be computed as the *cyclic* convolution of two length N sequences, obtained by appending $N - d$ and $N - k$ zeroes to the length k and length d sequences, respectively. Consequently, if $C(A\mathbf{z} \bullet B\mathbf{y})$ is a bilinear algorithm to compute the cylic convolution of two indeterminate sequences, z_i and y_i, $i = 0, 1, 2, \ldots, N-1$, then the algorithm to compute the *aperiodic* convolution of a length k sequence, z_i, $i = 0, 1, 2, \ldots, k-1$, and a length d sequence, y_j, $j = 0, 1, 2, \ldots, d-1$, can be written as $C(A'\mathbf{z} \bullet B'\mathbf{y})$, where the matrices A' and B' can be obtained from A

and B by deleting the last $N - k$ and last $N - d$ *consecutive* columns of A and B, respectively. It is clear from our discussion that in such a case, $(A')^T$ is the generator matrix of the associated linear $(n, k, \bar{d})$ code. An example follows to illustrate some of the above concepts.

EXAMPLE 2. Let $k = 4$ and $d = 4$; the system Ψ is given by

$$
\psi = \begin{bmatrix} x_0 & x_1 & x_2 & x_3 \\ x_1 & x_2 & x_3 & x_4 \\ x_2 & x_3 & x_4 & x_5 \\ x_3 & x_4 & x_5 & x_6 \end{bmatrix} \begin{bmatrix} y_0 \\ y_1 \\ y_2 \\ y_3 \end{bmatrix},
$$

where we may think of ψ_t, $t = 0, 1, \ldots, k - 1$, as the first four outputs of a simple 4-point FIR filter. Note that if we compute these values in a straightforward manner, 16 multiplications and 12 additions would be required. Taking a direct approach, for which *symbolic manipulation* software can be most helpful, the NC algorithm for computing this system, $\psi = C(A\mathbf{x} \bullet B\mathbf{y})$, may be verified to have matrices

$$
A = \begin{bmatrix} 1 & -1 & -1 & 1 & 0 & 0 & 0 \\ 0 & 1 & 0 & -1 & 0 & 0 & 0 \\ 0 & 1 & -1 & -1 & 1 & 0 & 0 \\ 0 & 0 & 1 & -1 & 0 & 0 & 0 \\ 0 & 0 & 0 & 1 & 0 & 0 & 0 \\ 0 & 0 & 0 & 1 & -1 & 0 & 0 \\ 0 & 0 & 1 & -1 & -1 & 1 & 0 \\ 0 & 0 & 0 & 1 & 0 & -1 & 0 \\ 0 & 0 & 0 & 1 & -1 & -1 & 1 \end{bmatrix}, \quad
B = \begin{bmatrix} 1 & 0 & 0 & 0 \\ 1 & 1 & 0 & 0 \\ 0 & 1 & 0 & 0 \\ 1 & 0 & 1 & 0 \\ 1 & 1 & 1 & 1 \\ 0 & 1 & 0 & 1 \\ 0 & 0 & 1 & 0 \\ 0 & 0 & 1 & 1 \\ 0 & 0 & 0 & 1 \end{bmatrix} \quad \text{and}
$$

$$
C = \begin{bmatrix} 1 & 1 & 0 & 1 & 1 & 0 & 0 & 0 & 0 \\ 0 & 1 & -1 & 0 & 1 & -1 & 0 & 0 & 0 \\ 0 & 0 & 0 & 1 & 1 & 0 & -1 & -1 & 0 \\ 0 & 0 & 0 & 0 & 1 & -1 & 0 & -1 & 1 \end{bmatrix}.
$$

This algorithm may also be easily obtained by partitioning X into 2×2 blocks and computing Ψ using three smaller (2×2)-dimensional bilinear systems (all having a matrix of indeterminates with the same structure as X), each requiring $\mu = 3$ multiplications (and 4 additions). The complexity for the above problem is then $n = 9$ multiplications (and 20 additions) over $\mathcal{F}$. We note that $n > k + d - 1 = 7$.

Assume that $\mathcal{F} = GF(2)$, the Galois field having two elements. We may then regard C as the generator matrix of the linear $(9, 4, 4)$ code generated over $GF(2)$ (actually, a code family is generated of the form (n, k, d), where $k + d - 1 = $ constant). Note that entries -1 in C are equivalent to 1 over $GF(2)$. It is interesting to consider the smaller two-point computation referred to earlier in this example and of the form

$$\psi^{(2)} = \begin{bmatrix} x_0 & x_1 \\ x_1 & x_2 \end{bmatrix} \begin{bmatrix} y_0 \\ y_1 \end{bmatrix}.$$

Call $C^{(2)}$ the generator matrix of the associated $(3, 2, 2)$ linear code; then, we find that $C^{(2)}$ is embedded in C in the following manner:

$$C = \begin{bmatrix} C^{(2)} & C^{(2)} & 0 \\ 0 & C^{(2)} & C^{(2)} \end{bmatrix}.$$

Also notice that $B^T = C$ over $GF(2)$ for this example.

The A-dual Φ of the system Ψ is given by

$$\phi = \begin{bmatrix} z_0 & 0 & 0 & 0 \\ z_1 & z_0 & 0 & 0 \\ z_2 & z_1 & z_0 & 0 \\ z_3 & z_2 & z_1 & z_0 \\ 0 & z_3 & z_2 & z_1 \\ 0 & 0 & z_3 & z_2 \\ 0 & 0 & 0 & z_3 \end{bmatrix} \begin{bmatrix} y_0 \\ y_1 \\ y_2 \\ y_3 \end{bmatrix},$$

which is computable using the *NC* algorithm $A^T (C^T \mathbf{z} \bullet B\mathbf{y})$ with $n = 9$ multiplications. Using the generating polynomials,

$$Z(u) = z_0 + z_1 u + z_2 u^2 + z_3 u^3, \quad Y(u) = y_0 + y_1 u + y_2 u^2 + y_3 u^3,$$

the bilinear forms ψ_t, $t = 0, 1, \ldots, N - 1$, where $N = k + d - 1 = 7$, are coefficients of the polynomial resulting from the product

$$
\begin{aligned}
\Phi(u) &= Z(u)Y(u) \\
&= z_0 y_0 + (z_1 y_0 + z_0 y_1)u + (z_2 y_0 + z_1 y_1 + z_0 y_2)u^2 \\
&\quad + (z_3 y_0 + z_2 y_1 + z_1 y_2 + z_0 y_3)u^3 + (z_3 y_1 + z_2 y_2 + z_1 y_3)u^4 \\
&\quad + (z_3 y_2 + z_2 y_3)u^5 + z_3 y_3 u^6,
\end{aligned}
$$

which has degree $N - 1 = k + d - 2$.

Using (2.15) in addition to the property that over the field $GF(p^m)$ having at least $N - 1$ elements, the multiplicative complexity of an aperiodic convolution of length N expressed as the polynomial product $Z(u)Y(u)$ is N, Reed-Solomon (RS) codes were constructed in [2.8]. The authors also constructed Bose-Chaudhary-Hocquenghem (BCH) codes from a transformation of the system (2.15).

In the next chapter, we take up the task of deriving efficient cyclic and aperiodic convolution algorithms over finite fields. The former algorithms will prove useful in applications involving filtering, Fourier transformation, and coding theory, as regards the first approach to code construction. The latter algorithms will prove useful in applications involving filtering, polynomial and larger integer products, and coding theory, as regards the second approach to code construction.

2.5   References

[**2.1**] A. V. Aho, J. E. Hopcroft, and J. D. Ullman, *The Design and Analysis of Computer Algorithms*, Addison-Wesley, 1974.

[**2.2**] S. Winograd, "On the Number of Multiplications Necessary to Compute Certain Functions," *Comm. Pure and Appl. Math.*, vol. 23, pp. 165-179, 1970.

[**2.3**] C. M. Fiduccia, "Fast Matrix Multiplication," in *Proc. Third Ann. ACM Symp. on Theory of Computing*, Shaker Heights, Ohio, pp. 45-49, May 1971.

[**2.4**] J. Hopcroft and J. Musinski, "Duality Applied to the Complexity of Matrix Multiplication," *SIAM Journ. Comput.*, vol. 2, pp. 159-173, Sept. 1973.

[**2.5**] C. M. Fiduccia, *On Obtaining Upper Bounds on the Complexity of Matrix Multiplication*, Plenum Press, New York, 1972.

[**2.6**] V. Strassen, *Evaluation of Rational Functions*, Plenum Press, New York, 1972.

[2.7] D. Dobkin, "On the Arithmetic Complexity of a Class of Computations," Thesis in the Division of Engineering and Applied Physics, Harvard University, Cambridge, MA, Sept. 1973.

[2.8] A. Lempel and S. Winograd, "A New Approach to Error-Correcting Codes," *IEEE Trans. Inform. Theory,* vol. IT-23, pp. 503-508, July 1977.

[2.9] R. W. Brockett and D. Dobkin, "On the Optimal Evaluation of a Set of Bilinear Forms," in Proc. *Fifth Ann. ACM Symp. on Theory of Computing,* Austin, Texas, pp. 88-95, April-May 1973.

[2.10] F. J. MacWilliams and N. J. A. Sloane, *The Theory of Error-Correcting Codes,* North Holland, 1977.

Chapter 3

Efficient Finite Field Algorithms

The discrete convolution of sequences plays an important role in signal processing. Aperiodic convolution is used for predicting system responses to nonperiodic inputs and for computing polynomial and large integer products. Periodic, or cyclic, convolution is used for predicting the output of a linear-time-invariant finite impulse response (FIR) system to a periodic input and can be employed to compute the discrete Fourier transform (DFT). It is also possible to *embed* a problem requiring aperiodic convolution in a cyclic convolution problem; thus, in this chapter, we first treat cyclic convolution algorithms.

There is a large body of literature on the discrete convolution of sequences, largely over the real and complex number fields [3.1 - 3.14]; however, the discrete convolution of sequences over finite fields, although equally important, has received far less attention. Early approaches to computational algorithms for cyclic convolution were based principally on the Chinese Remainder Theorem (CRT) [3.15], i.e., the breaking up of the original problem length into a number of smaller problems having mutually prime lengths for which efficient cyclic convolution algorithms were available. Some of the earliest work in the engineering literature using this approach may be found in [3.1].

Two disadvantages with this approach are that the convolution lengths that can be reconstructed from the product of mutually prime factors are somewhat limited and that algorithms obtained in this fashion have little regular structure which may be exploited in hardware or software design. Nonetheless, these algorithms are indeed interesting, quite efficient, and useful in a range of applications. Efforts to introduce regularity into the algorithms led to the discovery of polynomial transforms [3.13 - 3.14, 3.16]. Though algorithms obtained in this manner are more structured than those of [3.1], they still lack a completely regular structure.

The first half of this chapter takes a new approach to the design of length N cyclic convolution bilinear algorithms over $\mathcal{F} = GF(p^m)$, where p is a prime and m is an arbitrary positive integer. When multiplicative complexity is the prime computational issue, the method presented permits *systematic, structured* design of highly efficient algorithms. More details on the method employed may be found in [3.17 - 3.18]. The second half of this chapter deals with the efficient design of aperiodic convolution algorithms over $\mathcal{F}$. In this case, the CRT will be central to our development of aperiodic convolution algorithms. Additional details on the strategy employed may be found in [3.19 -3.20].

3.1   Cyclic Convolution

Before proceeding there are two results which are important. They are presented in the form of a lemma and a definition. We assume all functions to be computed are elements of $\mathcal{E} = \mathcal{F}(I)$, where $\mathcal{F} = GF(p^m)$, p prime. The elements of the set of indeterminates I may be referred to as the algorithm inputs.

LEMMA 3.1 *A bilinear cyclic convolution algorithm $\psi = C(Ax \bullet By)$ with field of constants $GF(p)$ which is valid for input data over $GF(p)$ is also valid for input data over $GF(p^m)$.*

PROOF. Let $\psi = x * y = C(Ax \bullet By)$, where x and y have elements from $GF(p)$ and A, B, and C are matrices of appropriate dimension over $GF(p)$. We prove that ψ can be computed in the same manner even if x and y are vectors over $GF(p^m)$. In this case, we can express x and y as

$$x = \sum_{i=0}^{m-1} x_i \alpha^i \text{ and } y = \sum_{j=0}^{m-1} y_j \alpha^j,$$

where $\mathbf{x}_i$ and $\mathbf{y}_j$ are vectors over $GF(p)$ and α is a primitive element of $GF(p^m)$. Using bilinearity, we obtain

$$\psi = (\sum_{i=0}^{m-1} \mathbf{x}_i \alpha^i) * (\sum_{j=0}^{m-1} \mathbf{y}_j \alpha^j) = \sum_{i=0}^{m-1} \sum_{j=0}^{m-1} \alpha^{i+j} (\mathbf{x}_i * \mathbf{y}_j).$$

Further, since $\mathbf{x}_i$ and $\mathbf{y}_j$ are over $GF(p)$, we may use the bilinear algorithm to compute $\mathbf{x}_i * \mathbf{y}_j$; thus,

$$
\begin{aligned}
\psi &= \sum_{i=0}^{m-1} \sum_{j=0}^{m-1} \alpha^{i+j} [C(A\mathbf{x}_i \bullet B\mathbf{y}_j)] \\
&= C \left[\sum_{i=0}^{m-1} \sum_{j=0}^{m-1} (A\mathbf{x}_i \alpha^i \bullet B\mathbf{y}_j \alpha^j) \right] \\
&= C \left[(A \sum_{i=0}^{m-1} \mathbf{x}_i \alpha^i) \bullet (B \sum_{j=0}^{m-1} \mathbf{y}_j \alpha^j) \right] = C(A\mathbf{x} \bullet B\mathbf{y}).
\end{aligned}
$$

This result is useful, as it allows us to confine consideration to a smaller field of constants, $GF(p)$. We note that this result is specialized to the operation of cyclic convolution, but that, in general, NC algorithms are developed on the premise that indeterminates do not commute and, thus, matrices may be substituted for scalar indeterminates without affecting the validity of the algorithm.

DEFINITION 1. Let $N = p^n - 1$ for some integer n, and partition the integers $[0, N - 1]$ into the sets $S_{i_1}, S_{i_2}, \ldots$. A *cyclotomic set* (conjugate class) S_i is generated by starting from the smallest integer i not covered in an earlier set and constructing other members as

$$S_i = \{i, ip \bmod N, ip^2 \bmod N, \ldots\}.$$

Since $ip^n \bmod N = i$, each set S_i is finite. Let the cardinality $| S_i | = \sigma_i$ and denote the set of indices $\{i_1, i_2, \ldots, \}$ by S. Some properties of the σ_i are:

P1) $0 \epsilon S$ and $\sigma_0 = 1$.

P2) Given any divisor t of n (including $t = 1$ when $p \neq 2$), $\theta = (p^n - 1)/(p^t - 1) \epsilon S$ and $\sigma_\theta = t$.

P3) For any nonzero $i \epsilon S$, it is possible to find a $\theta \epsilon S$ having the form given in P2 for some t, such that i is a multiple of θ and $\sigma_i = \sigma_\theta$.

P4) With θ as in P2 and α, a primitive element of $GF(p^n)$, α^θ generates the subfield $GF(p^{\sigma_\theta})$ of $GF(p^n)$ and, consequently, $\sigma_\theta \mid n$.

P5) From P3 and P4, for any $i \epsilon S$, $\sigma_i \mid n$ and $\alpha^i \epsilon GF(p^{\sigma_i})$.

Proofs of the properties associated with cyclotomic sets may be found in Appendix A. The following example illustrates how such sets are constructed; the reader may also wish to refer to [2.10].

EXAMPLE 1.   Consider $GF(2^4)$ and $N = 2^4 - 1 = 15$ ($n = m = 4$). We have

$$S_0 = \{0\} \qquad\qquad\quad \sigma_0 = 1,$$
$$S_1 = \{1, 2, 4, 8\} \qquad\quad \sigma_1 = 4,$$
$$S_3 = \{3, 6, 12, 9\} \qquad\;\, \sigma_3 = 4,$$
$$S_5 = \{5, 10\} \qquad\qquad\; \sigma_5 = 2,$$
$$S_7 = \{7, 14, 13, 11\} \qquad \sigma_7 = 4, \text{ and}$$
$$S = \{0, 1, 3, 5, 7\} \;.$$

Note in this example that $n = 4$ has only two divisors t, 2 and 4. From P2, the values of θ corresponding to these divisors are 5 and 1, respectively. Both of these integers belong to S. The other nonzero members, 3 and 7, of S are both multiples of 1 and $\sigma_3 = \sigma_7 = \sigma_1$, in accord with P3.

Let $S = S_N \cup \{0\}$. The reader should recognize that the above partitioning corresponds to a partitioning of the elements of $GF(p^n)$, where the integers in any set S_i are the exponents of α, a primitive element in $GF(p^n)$. Each cyclotomic set S_i, $i \epsilon S_N$, is a grouping of elements of *order* N_i/N, where $N_i = (p^n - 1)/\gcd(i, p^n - 1)$, having a *multiplicative order* of p mod N_i equal to $\sigma_i \mid n$. In Example 1, cyclotomic sets S_1 and S_7 correspond to elements of order 15 and each has multiplicative order 4; in this case, these eight elements comprise all the primitive elements in $GF(2^4)$. The cyclotomic sets S_3 and S_5 correspond to elements of order 5 and 3, respectively. Each set S_i also corresponds to an irreducible polynomial $P_i(u)$, the *minimal polynomial* of α^i over $GF(p)$. For Example 1, it is easy to verify that $P_1(u) = u^4 + u + 1$, $P_3(u) = u^4 + u^3 + u^2 + u + 1$, $P_5(u) = u^2 + u + 1$, $P_7(u) = u^4 + u^3 + 1$; the remaining polynomial, $P_0(u) = u - 1$, always corresponds to the cyclotomic set S_0. We leave it to the reader to verify that the product of these (monic) distinct, irreducible polynomials over $GF(p)$ is the polynomial $u^{15} - 1$. The field $GF(2^4)$ is sometimes called the *splitting field* for $u^{15} - 1$ (or $u^{16} - u$) over $GF(p)$. In Example 1, $GF(p^{\sigma_5}) = GF(2^2) \subset GF(2^4)$ is a *subfield* of the splitting field and a splitting field itself for $u^3 - 1$ (or $u^4 - u$) over $GF(p)$.

3.1.1 Algorithm Development

As noted in Chapter 2, the cyclic convolution of sequences **x** and **y**, each of length N, is given by

$$\psi_t = \sum_{i=0}^{N-1} x_i y_{t-i}, \qquad t = 0, 1, \ldots, N - 1,$$

where the subscript $t - i$ is understood to mean $(t - i) \bmod N$. We represent the resulting length N sequence as $\psi = \mathbf{x} * \mathbf{y}$. In this section, we design a bilinear algorithm to compute ψ for data sequences over $\mathcal{F} = GF(p^m)$, a field of constants $GF(p)$, and $N = p^n - 1$ (this does somewhat restrict the possible cyclic convolution lengths; however, we shall see in the sequel that it is possible to design cyclic convolution algorithms of *any* length N relatively prime to p). From the above lemma, it is sufficient to assume that the components of **x** and **y** are from $GF(p)$.

Now, it is well known that ψ can also be obtained from **X** and **Y**, the Fourier transforms of **x** and **y** in $GF(p^n)$, as

$$\psi_t = (1/N) \sum_{j=0}^{N-1} X_j Y_j \, \alpha^{-jt}, \qquad t = 0, 1, \ldots, N - 1, \qquad (3.1)$$

where α is a primitive element of $GF(p^n)$. As in [3.17], we first carry out the summation (3.1) over the σ_i elements of each cyclotomic set S_i and then over all such sets, i.e.,

$$\psi_t = \sum_{i \in S} \psi_{it}, \tag{3.2}$$

where

$$\psi_{it} = (1/N) \sum_{j \in S_i} X_j Y_j \, \alpha^{-jt}, \qquad t = 0, 1, \ldots, N-1. \tag{3.3}$$

Define $\boldsymbol{\psi}_i = [\psi_{io} \ \psi_{i1} \cdots \psi_{i,N-1}]^T$. We now show that $\boldsymbol{\psi}_i$ can be computed by an NC algorithm of the form $C(Ax \bullet By)$.

THEOREM 3.1 *The vector* $\boldsymbol{\psi}_i$ *can be computed from the* **x** *and* **y** *vectors by a bilinear algorithm over* $GF(p)$.

PROOF. Let $i = 0$. The elements ψ_{ot} are given by

$$\psi_{ot} = (1/N) \left(\sum_{i=0}^{N-1} x_i \right) \left(\sum_{j=0}^{N-1} y_j \right), \qquad t = 0, 1, \ldots, N-1;$$

thus,

$$\boldsymbol{\psi}_0 = \mathbf{C}_0(\mathbf{A}_0\mathbf{x} \bullet \mathbf{B}_0\mathbf{y}),$$

where $\mathbf{A}_0$, $\mathbf{B}_0$, and $\mathbf{C}_0$ are the length N vectors

$$\mathbf{A}_0 = [11 \cdots 1], \ \mathbf{B}_0 = \mathbf{A}_0, \text{ and}$$
$$\mathbf{C}_0 = (1/N)\mathbf{A}_0^T.$$

For $i \neq 0$, since any integer $j \in S_i$ can be expressed as $ip^\ell \bmod N$, $\ell = 0, 1, \ldots, \sigma_i - 1$, we may write the Fourier coefficients corresponding to S_i as

$$X_j = \sum_{k=0}^{N-1} x_k \alpha^{kip^\ell}, \ j = ip^\ell; \ \ell = 0, 1, \ldots, \sigma_i - 1.$$

The fact that $x_k \in GF(p)$ implies that $(x_k)^{p^\ell} = x_k$; therefore, for $j = ip^\ell$ we have that

$$X_j = \sum_{k=0}^{N-1} \left(x_k \alpha^{ki} \right)^{p^\ell} = \left(\sum_{k=0}^{N-1} x_k \alpha^{ki} \right)^{p^\ell} = (X_i)^{p^\ell}.$$

In a similar manner, we have $Y_j = (Y_i)^{p^\ell}$. Using these results, the elements of ψ_i are given by

$$\psi_{it} = (1/N) \sum_{\ell=0}^{\sigma_i-1} (X_i Y_i \alpha^{-it})^{p^\ell}.$$

We examine this expression more closely. Note that $X_i = \sum_{k=0}^{N-1} x_k \alpha^{ik}$, where i is a multiple of some $\theta = (p^n - 1)/(p^{\sigma_\theta} - 1) \epsilon S$ by $P2$ and $P3$. This implies that X_i is a polynomial in α^θ. From $P4$, the minimal polynomial over $GF(p)$ of α^θ is of degree σ_θ. Using this minimal polynomial to reduce the σ_θth and higher powers in α^θ, X_i can be reduced to a polynomial in α^θ of degree $\sigma_\theta - 1$. Each coefficient of this polynomial is obtained by a linear combination over $GF(p)$ of the components of $\mathbf{x}$. The same is true with respect to Y_i and $\mathbf{y}$. The matrices A and B are thus established. We may now use bilinear small degree polynomial multiplication algorithms over $GF(p)$ to obtain the coefficients r_m of the product polynomial $X_i Y_i$ in α^θ of degree $2\sigma_\theta - 2$. To prove the theorem, it remains to show that ψ_{it} for any t can be expressed as a linear combination of the r_m coefficients over $GF(p)$. Using the above form for ψ_{it}, we obtain

$$
\begin{aligned}
\psi_{it} &= (1/N) \sum_{\ell=0}^{\sigma_i-1} \left(\left(\sum_{m=0}^{2\sigma_\theta-2} r_m \alpha^{\theta m} \right) \alpha^{-it} \right)^{p^\ell} \\
&= (1/N) \sum_{\ell=0}^{\sigma_i-1} \sum_{m=0}^{2\sigma_\theta-2} r_m (\alpha^{\theta m - it})^{p^\ell} \\
&= (1/N) \sum_{m=0}^{2\sigma_\theta-2} r_m tr(\alpha^{\theta m - it}),
\end{aligned}
\tag{3.4}
$$

where the trace function, $tr(\cdot) : GF(p^{\sigma_\theta}) \to GF(p)$, is defined [3.21] as,

$$tr(\beta) = \sum_{\ell=0}^{\sigma_\theta-1} \beta^{p^\ell}, \qquad \beta \epsilon GF(p^{\sigma_\theta}).$$

We see, therefore, that ψ_{it} is always a linear combination over $GF(p)$ of the coefficients r_m, $m = 0, 1, \ldots, 2\sigma_\theta - 2$, and the matrix C is thus established.

Theorem 3.1 asserts that there exist matrices A_i, B_i, and C_i over $GF(p)$ such that

$$\psi_i = C_i(A_i\mathbf{x} \bullet B_i\mathbf{y}), \qquad i\epsilon S. \qquad (3.5)$$

From (3.2), the bilinear algorithm over $GF(p)$ for the overall problem is then immediately obtained as

$$\psi = C(A\mathbf{x} \bullet B\mathbf{y}), \qquad (3.6)$$

where

$$C = [C_{i_1}, C_{i_2}, \cdots], \quad A = [A_{i_1}^T, A_{i_2}^T, \cdots]^T, \text{ and}$$

$$B = [B_{i_1}^T, B_{i_2}^T, \cdots]^T, \qquad (3.7)$$

with $i_1, i_2, \ldots \epsilon S$.

We are now faced with the important problem of how to construct the matrices A_i, B_i, and C_i for all $i\epsilon S$. For $i = 0$, this is trivial (see the proof of Theorem 3.1). For $i \neq 0$, we see from $P3$ that it is possible to find a $\theta = (p^n - 1)/(p^{\sigma_\theta} - 1)\epsilon S$ such that i is a multiple of θ and $\sigma_i = \sigma_\theta$. The following theorem assures us that, for each such i and θ, the matrices A_i, B_i, and C_i are related to the matrices A_θ, B_θ, and C_θ.

THEOREM 3.2 *For integers* $i\epsilon S$ *and* $\theta\epsilon S$, *if* $\theta|i, \sigma_\theta = \sigma_i$ *and* $\theta = (p^n - 1)/(p^{\sigma_\theta} - 1)$, *then*

$$A_i(\gamma, \delta) = A_\theta(\gamma, (i/\theta)\delta \bmod(N/\theta)),$$

$$B_i(\gamma, \delta) = B_\theta(\gamma, (i/\theta)\delta \bmod(N/\theta)),$$

$$C_i(\gamma, \delta) = C_\theta((i/\theta)\gamma \bmod(N/\theta), \delta),$$

where $A_i(\gamma, \delta)$ *denotes the* (γ, δ) *element of* A_i.

PROOF. The A_i and B_i matrices are used to obtain suitable linear forms in the elements of **x** and **y**, respectively, in order to effect multiplication of the $\sigma_\theta - 1$ degree polynomials X_i and Y_i in α^θ. Similarly, A_θ**x** and B_θ**y** provide the linear forms required for the multiplication of the $\sigma_\theta - 1$ degree polynomials X_θ and Y_θ in α^θ. Since, in both cases, the polynomial degrees are identical, identical multiplication algorithms could be employed. Note, however, that $X_i = \sum_{k=0}^{N-1} x_k \alpha^{ik}$ and $X_\theta = \sum_{k=0}^{N-1} x_k \alpha^{\theta k}$. Suppose, in the latter expression, that the coefficient of α^{ik} is $x_{k'}$; then, we have $\theta k' = ik \bmod N$ or $k' = (i/\theta)k \bmod(N/\theta)$. Thus, if x_k is replaced by $x_{k'}$, where $k' = (i/\theta)k \bmod(N/\theta)$, in the polynomial X_i, we obtain the polynomial X_θ. The relation between A_i and A_θ immediately follows; identical arguments applied to Y_i and Y_θ result in the relation between B_i and B_θ.

Starting now from the relation (3.4), i.e.,

$$\psi_{it} = (1/N) \sum_{m=0}^{2\sigma_\theta - 2} r_m \, tr(\alpha^{\theta m - it}),$$

we have

$$\psi_{\theta t'} = (1/N) \sum_{m=0}^{2\sigma_\theta - 2} r_m \, tr\left(\alpha^{\theta m - \theta((i/\theta)\gamma \bmod(N/\theta))}\right)$$

$$= (1/N) \sum_{m=0}^{2\sigma_\theta - 2} r_m tr\left(\alpha^{\theta m - i\gamma}\right) = \psi_{i\gamma},$$

where $t' = (i/\theta)\gamma \bmod(N/\theta)$. This immediately provides the desired relation between C_i and C_θ.

From Theorem 3.2, it is only necessary that we obtain the matrices A_θ, B_θ, and C_θ for all $\theta \epsilon S$ of the type $\theta = (p^n - 1)/(p^{\sigma_\theta} - 1)$. Note that for such a θ, α^θ is the root of a primitive polynomial of degree σ_θ over $GF(p)$. We call these the *fundamental transformation matrices* associated with the *NC* algorithm $C(A\mathbf{x} \bullet B\mathbf{y})$. To keep our ideas concrete, we again consider Example 1 for which $\theta = 1, 5$. We have α a primitive element of $GF(2^4)$ with primitive polynomial $P_1(u)$ of degree $\sigma_1 = 4$. Let $\beta = \alpha^5$ be a primitive element of $GF(2^2)$ (note that $\beta^2 + \beta + 1 = 0$); the corresponding primitive polynomial is $P_5(u)$

of degree $\sigma_5 = 2$. In this case, we have $\hat{P}_5(u) = P_5(u)$, where for an arbitrary polynomial $P(u)$ of degree l, the reciprocal polynomial is defined as $\hat{P}(u) = u^l P(1/u)$. For future reference, we also note that with regard to Example 1, another primitive polynomial is associated with the primitive element α^7 of $GF(2^4)$ and that $P_7(u) = \hat{P}_1(u)$. The following definition will assist us in identifying a very important structure inherent in the fundamental matrices A_θ, B_θ, and C_θ.

DEFINITION 2.   Consider the primitive polynomial of degree σ_θ

$$P_\theta(u) = u^{\sigma_\theta} - a_1 u^{\sigma_\theta - 1} - a_2 u^{\sigma_\theta - 2} - \cdots - a_{\sigma_\theta}$$

in the indeterminate u having coefficients from $GF(p)$. A *linear periodic recurrent sequence* $\{u_i\}$ of elements from $GF(p)$ with period $p^{\sigma_\theta} - 1$ can be obtained from the difference equation

$$u_i = a_1 u_{i-1} + a_2 u_{i-2} + \cdots + a_{\sigma_\theta} u_{i-\sigma_\theta}$$

with arbitrary, but nonzero, initial conditions. This sequence is also known as a *maximal length recurrent sequence* (MLRS) and exhibits pseudorandom properties [3.22]. We refer to it as an MLRS generated by $P_\theta(u)$. The primitive reciprocal polynomial, denoted $\hat{P}_\theta(u)$, and also of degree σ_θ, can be expressed as

$$\hat{P}_\theta(u) = u^{\sigma_\theta} - a_1' u^{\sigma_\theta - 1} - a_2' u^{\sigma_\theta - 2} - \cdots - a_{\sigma_\theta}'$$

in the indeterminate u having coefficients from $GF(p)$. The MLRS generated by $\hat{P}_\theta(u)$ can be obtained from the difference equation

$$u_i = a_1' u_{i-1} + a_2' u_{i-2} + \cdots + a_{\sigma_\theta}' u_{i-\sigma_\theta}$$

with arbitrary, but nonzero, initial conditions. Note that the primitive polynomials $P_\theta(u)$ and $\hat{P}_\theta(u)$ satisfy $P_\theta(\alpha^\theta) = \hat{P}_\theta(\alpha^{-\theta}) = 0$.

The next result exposes the connection between MLRS's and the fundamental transformation matrices A_θ, B_θ, and C_θ.

THEOREM 3.3 *The rows of A_θ and B_θ are MLRS's generated by the primitive polynomial $P_\theta(u)$, and the columns of C_θ are MLRS's generated by the primitive reciprocal polynomial $\hat{P}_\theta(u)$.*

PROOF. Consider the polynomial

$$X_\theta = \sum_{k=0}^{N-1} x_k \alpha^{\theta k} = \sum_{j=0}^{\sigma_\theta - 1} f_j(\alpha^\theta)^j,$$

where the expression involving the coefficients $f_j, j = 0, 1, \cdots, \sigma_\theta - 1$, follows from the fact that α^θ is a root of $P_\theta(u)$. It is not difficult to show that the coefficients $f_j, j = 0, 1, \cdots, \sigma_\theta - 1$, have the form

$$f_j = \sum_{k=0}^{N-1} \mathcal{E}_{jk} x_k, \qquad \mathcal{E}_{jk} \in GF(p).$$

We first show that the vector $\mathcal{E}_j = [\mathcal{E}_{jo} \, \mathcal{E}_{j1} \cdots \mathcal{E}_{j,N-1}]^T$ has components which is an MLRS generated by $P_\theta(u)$. From the above two expressions, we have

$$\sum_{k=0}^{N-1} x_k \alpha^{\theta k} = \sum_{k=0}^{N-1} x_k \sum_{j=0}^{\sigma_\theta - 1} \mathcal{E}_{jk} (\alpha^\theta)^j,$$

which implies

$$\alpha^{\theta k} = \sum_{j=0}^{\sigma_\theta - 1} \mathcal{E}_{jk} \alpha^{\theta j}.$$

Since $P_\theta(\alpha^\theta) = 0$, we also obtain, for any integer $k \geq \sigma_\theta$, that

$$\alpha^{\theta k} = \sum_{\ell=1}^{\sigma_\theta} a_\ell (\alpha^\theta)^{k-\ell},$$

where $a_\ell \in GF(p)$ is the coefficient of $u^{\sigma_\theta - \ell}$ in $P_\theta(u)$. Equating these expressions yields

$$\mathcal{E}_{jk} = \sum_{\ell=1}^{\sigma_\theta} a_\ell \mathcal{E}_{j,k-\ell},$$

from which it is clear that $\mathcal{E}_j$ is an MLRS generated by $P_\theta(u)$. Recall that $A_\theta \mathbf{x}$ gives the linear forms in the f_j coefficients used in the product $X_\theta Y_\theta$. Each row of A_θ is a linear combination over $GF(p)$ of the vector $\mathcal{E}_j$. Since a linear combination of MLRS's generated by the same polynomial is again an MLRS [3.22], it follows that each row of A_θ is an MLRS generated by $P_\theta(u)$. The proof for B_θ is similar.

To prove the statement involving C_θ, we start with the result (3.4) i.e.,

$$\psi_{\theta t} = (1/N) \sum_{m=0}^{2\sigma_\theta - 2} r_m \, tr(\alpha^{\theta m - \theta t}).$$

Using the expression for the reciprocal polynomial $\hat{P}_\theta(u)$ and the fact that $\hat{P}_\theta(\alpha^{-\theta}) = 0$, we have

$$\alpha^{-\theta \sigma_\theta} = \sum_{\ell=1}^{\sigma_\theta} a'_\ell \, \alpha^{-\theta(\sigma_\theta - \ell)} \quad \text{or} \quad \sum_{\ell=1}^{\sigma_\theta} a'_\ell \, \alpha^{\theta \ell} = 1.$$

Now consider the following expression for any $t \geq \sigma_\theta$:

$$\sum_{\ell=1}^{\sigma_\theta} a'_\ell \psi_{\theta, t-\ell} = (1/N) \sum_{\ell=1}^{\sigma_\theta} a'_\ell \sum_{m=0}^{2\sigma_\theta - 2} r_m \, tr(\alpha^{\theta m - \theta(t-\ell)}).$$

Using the linearity of $tr(\cdot)$ and the above identity obtained from $\hat{P}_\theta(u)$, this expression simplifies to

$$
\begin{aligned}
\sum_{\ell=1}^{\sigma_\theta} a'_\ell \, \psi_{\theta, t-\ell} &= (1/N) \sum_{m=0}^{2\sigma_\theta - 2} r_m \, tr \left(\sum_{\ell=1}^{\sigma_\theta} a'_\ell \, \alpha^{-\theta(t-m-\ell)} \right) \\
&= (1/N) \sum_{m=0}^{2\sigma_\theta - 2} r_m \, tr(\alpha^{-\theta(t-m)}) = \psi_{\theta t},
\end{aligned}
$$

which immediately shows that the columns of C_θ are MLRS's generated by $\hat{P}_\theta(u)$.

The results of Theorems 3.2 and 3.3 provide a structured design method for cyclic convolutions over finite fields [3.17]. In the next section, we list the steps involved in the algorithm and then provide an example which we first encountered in Chapter 2.

3.1.2 Algorithm Steps

The algorithm, of course, is data independent. If we consider a particular data sequence, $x_k = 0$, $k \geq \sigma_\theta$, we can express the coefficients f_k of Theorem 3.3 as $f_k = x_k$, $k = 0, 1, \cdots, \sigma_\theta - 1$; thus, the linear forms in the f_k's involved in the multiplication of polynomials of degree $\sigma_\theta - 1$ become the linear forms in the x_k's, $k = 0, 1, \cdots, \sigma_\theta - 1$. This leads us to the first step. **Steps 1-5** are carried out for all $\theta = (p^n - 1)/(p^{\sigma_\theta} - 1) \epsilon S$.

Step 1: Using a bilinear small degree polynomial multiplication algorithm for polynomials of degree $\sigma_\theta - 1$, directly obtain the first σ_θ elements of any row of A_θ.

Step 2: Extend the remaining portion of each row of A_θ by the MLRS generated by $P_\theta(u)$. The σ_θ elements obtained in **Step 1** provide the necessary initial conditions.

Step 3: Use the same procedure of **Steps 1 and 2** to obtain B_θ.

Step 4: Evaluate

$$\psi_{\theta t} = (1/N) \sum_{m=0}^{2\sigma_\theta - 2} r_m \, tr(\alpha^{-\theta(t-m)}), \qquad t = 0, 1, \cdots, \sigma_\theta - 1,$$

and then use a bilinear small degree polynomial multiplication algorithm to obtain the first σ_θ rows of C_θ.

Step 5: Complete the columns of C_θ by the MLRS generated by $\hat{P}_\theta(u)$. Note that the MLRS generated by $\hat{P}_\theta(u)$ can be obtained simply by reversing the MLRS generated by $P_\theta(u)$.

Step 6: Obtain all other matrices A_i, B_i, and C_i, $i \epsilon S$ and $i \neq \theta$, by the sampling procedure described in Theorem 3.2.

Step 7: Compute the cyclic convolution of two sequences of length $N = p^n - 1$ as $\psi = C(Ax \bullet By)$, where A, B, and C are as defined in (3.7).

Steps 1, 3, and 4 require efficient, bilinear small degree polynomial multiplication algorithms over $GF(p)$. Appendix B provides such algorithms for degrees ranging from zero to three and are valid over any field.

EXAMPLE 2. Let $n = m = 3$; thus, the finite field is $GF(2^3)$ and the objective is to design a cyclic convolution algorithm for two length $N = p^n - 1 = 7$ sequences over $GF(2^3)$.

The system of bilinear forms that we wish to compute is $\psi = \mathbf{x} * \mathbf{y}$, i.e.,

$$\psi = \begin{bmatrix} x_0 & x_6 & x_5 & x_4 & x_3 & x_2 & x_1 \\ x_1 & x_0 & x_6 & x_5 & x_4 & x_3 & x_2 \\ x_2 & x_1 & x_0 & x_6 & x_5 & x_4 & x_3 \\ x_3 & x_2 & x_1 & x_0 & x_6 & x_5 & x_4 \\ x_4 & x_3 & x_2 & x_1 & x_0 & x_6 & x_5 \\ x_5 & x_4 & x_3 & x_2 & x_1 & x_0 & x_6 \\ x_6 & x_5 & x_4 & x_3 & x_2 & x_1 & x_0 \end{bmatrix} \begin{bmatrix} y_0 \\ y_1 \\ y_2 \\ y_3 \\ y_4 \\ y_5 \\ y_6 \end{bmatrix}.$$

For this example, the cyclotomic sets and orders are $S_0 = \{0\}$, $\sigma_0 = 1$; $S_1 = \{1, 2, 4\}$, $\sigma_1 = 3$; and $S_3 = \{3, 6, 5\}$, $\sigma_3 = 3$. The index set $S = \{0, 1, 3\}$. Let α denote the primitive element of $GF(2^3)$ satisfying $\alpha^3 + \alpha + 1 = 0$. The only divisor, t, of n in this case is 3, excluding $t = 1$, since $p = 2$; thus, $\theta = (p^n - 1)/(p^t - 1) = 1$. For $\theta = 1$, the primitive polynomial is $P_1(u) = (u - \alpha)(u - \alpha^2)(u - \alpha^4) = u^3 + u + 1$. Since $\sigma_1 = 3$, we now refer to the polynomial multiplication algorithm for degree $\sigma_1 - 1 = 2$ in Appendix B. Using the linear forms in the f_k's, we proceed in the following manner to complete the first $\sigma_1 = 3$ elements of each row of A_1.

Step 1:

$$A_1 = \begin{bmatrix} 1 & 0 & 0 \\ 0 & 1 & 0 \\ 0 & 0 & 1 \\ 1 & 1 & 0 \\ 0 & 1 & 1 \\ 1 & 0 & 1 \end{bmatrix} \cdots .$$

The six rows of A_1 correspond precisely to the six multiplications $m_0, m_1, \ldots, m_5$ defined in the degree-2 polynomial multiplication algorithm.

Step 2: From $P_1(u)$, we obtain the difference equation $u_i = u_{i-2} + u_{i-3}$, generating the associated MLRS with period $p^{\sigma_1} - 1 = 7$. For each row of A_1, the initial conditions are given by the first $\sigma_1 = 3$ elements of that row found in **Step 1**. The remaining elements are then filled in to obtain the (6×7)-dimensional matrix

$$
A_1 = \begin{bmatrix}
1 & 0 & 0 & 1 & 0 & 1 & 1 \\
0 & 1 & 0 & 1 & 1 & 1 & 0 \\
0 & 0 & 1 & 0 & 1 & 1 & 1 \\
1 & 1 & 0 & 0 & 1 & 0 & 1 \\
0 & 1 & 1 & 1 & 0 & 0 & 1 \\
1 & 0 & 1 & 1 & 1 & 0 & 0
\end{bmatrix}.
$$

Step 3: Once again, use the degree-2 polynomial multiplication algorithm of Appendix B, more specifically, the linear forms in the g_k's and the difference equation $u_i = u_{i-2} + u_{i-3}$, to obtain the (6×7)-dimensional matrix B_1. Note that the degree-2 algorithm is *symmetric* with respect to the f_k and g_k (x_k *and* y_k) sequences; thus, $B_1 = A_1$.

Step 4: Since $\theta = 1$ and $\sigma_1 = 3$, the expression for ψ_{1t} must be evaluated for $t = 0, 1, 2$ to obtain the first three rows of C_1. The r_m's in ψ_{1t} refer to the coefficients of the product polynomial in the degree-2 polynomial multiplication algorithm and can be expressed in terms of the multiplications m_i. We have

$$
\psi_{1t} = (1/7) \sum_{m=0}^{4} r_m tr(\alpha^{m-t}).
$$

The trace function can easily be shown to have the following values over $GF(2^3)$:

$$
tr(\alpha^i) = \begin{cases} 1, & i = 0, 3, 5, 6, 7; \\ 0, & i = 1, 2, 4. \end{cases}
$$

We also have $(1/7) = 1$ in $GF(2)$. Using these results, we obtain

$$
\begin{aligned}
\psi_{10} &= r_0 + r_3 = m_0 + m_1 + m_2 + m_4, \\
\psi_{11} &= r_0 + r_1 + r_4 = m_1 + m_2 + m_3, \ and \\
\psi_{12} &= r_0 + r_1 + r_2 = m_0 + m_2 + m_3 + m_5.
\end{aligned}
$$

From these linear forms, we obtain the first $\sigma_1 = 3$ rows of C_1 as

$$
C_1 = \begin{bmatrix}
1 & 0 & 1 \\
1 & 1 & 0 \\
1 & 1 & 1 \\
0 & 1 & 1 \\
1 & 0 & 0 \\
0 & 0 & 1
\end{bmatrix}^{T} \cdots \quad .
$$

Step 5: We extend each column of C_1 by the MLRS generated by the reciprocal polynomial $\hat{P}_1(u)$. Since $P_1(u) = u^3 + u + 1$, we have $\hat{P}_1(u) = u^3(u^{-3} + u^{-1} + 1) = u^3 + u^2 + 1$, for which we obtain the difference equation $u_i = u_{i-1} + u_{i-3}$. Using the elements of C_1 obtained in **Step 4** as initial conditions, we have

$$
C_1 = \begin{bmatrix}
1 & 0 & 1 & 0 & 0 & 1 & 1 \\
1 & 1 & 0 & 1 & 0 & 0 & 1 \\
1 & 1 & 1 & 0 & 1 & 0 & 0 \\
0 & 1 & 1 & 1 & 0 & 1 & 0 \\
1 & 0 & 0 & 1 & 1 & 1 & 0 \\
0 & 0 & 1 & 1 & 1 & 0 & 1
\end{bmatrix}^{T} .
$$

This completes our computation of the fundamental matrices A_1, B_1, and C_1.

Step 6: Sampling the rows of A_1 with a sampling period of 3, in accord with Theorem 3.2, we have

$$
A_3 = \begin{bmatrix}
1 & 1 & 1 & 0 & 1 & 0 & 0 \\
0 & 1 & 0 & 0 & 1 & 1 & 1 \\
0 & 0 & 1 & 1 & 1 & 0 & 1 \\
1 & 0 & 1 & 0 & 0 & 1 & 1 \\
0 & 1 & 1 & 1 & 0 & 1 & 0 \\
1 & 1 & 0 & 1 & 0 & 0 & 1
\end{bmatrix} .
$$

Sampling the columns of C_1 with a sampling period 3 leads to

$$C_3 = \begin{bmatrix} 1 & 0 & 1 & 1 & 1 & 0 & 0 \\ 1 & 1 & 1 & 0 & 0 & 1 & 0 \\ 1 & 0 & 0 & 1 & 0 & 1 & 1 \\ 0 & 1 & 0 & 1 & 1 & 1 & 0 \\ 1 & 1 & 0 & 0 & 1 & 0 & 1 \\ 0 & 1 & 1 & 1 & 0 & 0 & 1 \end{bmatrix}^T .$$

This completes the computation of A_i, B_i, and C_i for all $i \epsilon S$, other than $i = 0$.

Step 7: Construct the matrices

$$A = [\mathbf{A}_0^T, A_1^T, A_3^T]^T, \qquad B = A,$$

where $\mathbf{A}_0 = [11 \cdots 1]$, and

$$C = [\mathbf{C}_0, C_1, C_3],$$

where $\mathbf{C}_0 = [11 \cdots 1]^T = \mathbf{A}_0^T$. The cyclic convolution algorithm of two length $N = p^n - 1 = 7$ sequences over $GF(2)$ is then given by

$$\psi = C(A\mathbf{x} \bullet B\mathbf{y}).$$

In this example, we observe that A (and B) is of dimension (13×7); thus, the number of multiplications required to compute $A\mathbf{x} \bullet B\mathbf{y} = \mathbf{m}$ is 13 over $GF(2)$. The matrix C is of dimension (7×13).

Now consider Example 2 of Chapter 2, i.e., the computation of the system of bilinear forms (2.15) for $k = d = 4$. Let $y_{N-j} = v_j$, $j = 0, 1, \ldots, N - 1$, with $y_j = 0$, $j = d, \ldots, N - 1$, and $N = k + d - 1$; call the resulting vector $\mathbf{v}$. The algorithm developed in the above example can then compute the cyclic convolution (2.19) using $\psi' = C(A\mathbf{x} \bullet B\mathbf{v})$ from which we may extract the $k = 4$ solutions to (2.15). The linear (n, k, d) code generated over $GF(2)$ is the $(13,4,4)$ code. The generator matrix of the (13,4,4) code may easily be obtained by simply dropping any $(N - k)$ *consecutive* rows of C; taking the first 4 rows, we obtain the generator matrix as

$$
C' = \begin{bmatrix}
1 & 1 & 1 & 1 & 0 & 1 & 0 & 1 & 1 & 1 & 0 & 1 & 0 \\
1 & 0 & 1 & 1 & 1 & 0 & 0 & 0 & 1 & 0 & 1 & 1 & 1 \\
1 & 1 & 0 & 1 & 1 & 0 & 1 & 1 & 1 & 0 & 0 & 0 & 1 \\
1 & 0 & 1 & 0 & 1 & 1 & 1 & 1 & 0 & 1 & 1 & 0 & 1
\end{bmatrix}.
$$

Some optimization is possible since columns 3 and 8 (first column is column zero) are seen to be identical. The corresponding multiplications m_3 *and* m_8 could be evaluated as the single multiplication $m_3 + m_8 = (x_1 + x_2)(y_1 + y_2 + y_3)$. Thus, one of these columns can be dropped, and the multiplication associated with the remaining column can be replaced by $m_3 + m_8$. The result is the generator matrix for the $(12, 4, 4)$ linear code. The codeword length can be further shortened without changing the design minimum distance; however, we leave this subject for a later chapter.

The above design procedure for cyclic convolution is seen to be highly *systematic* and results in transformation matrices A, B, and C having a great deal of *inherent structure*. Large sizes can be dealt with relatively easily. An example of this is the length $N = 15$ algorithm over $GF(2)$ developed in [3.17]. The length $N = 15$ cyclic convolution algorithm has associated with it a $(31, 8, 8)$ linear code whose actual minimum distance is $\bar{d} = 9$.

3.1.3   Length p − 1 Algorithms Over a Field of Constants GF(p)

When $n = 1$, $N = p - 1$ and the field of constants is $GF(p)$. Due to the manner in which the cyclotomic sets are constructed, we have that for each i, $S_i = \{i\}$, $\sigma_i = 1$, and $S = \{0, 1, 2, \cdots, p - 2\}$. The matrices A_1 and B_1 are $(1 \times (p - 1))$-dimensional vectors whose first elements are unity, with the remaining elements filled in by the MLRS over $GF(p)$ of period $(p - 1)$ generated by $P_1(u)$. Other A_i and B_i, $i \neq 1$, are obtained by sampling the rows of A_1 and B_1, respectively, using a sampling period i. In this way, the matrices A and B are obtained. The matrix C is constructed in a similar fashion using the reciprocal polynomial $\hat{P}_1(u)$. Note that we simply have $\psi_{10} = r_0 = f_0 \cdot g_0$ from the degree zero algorithm of Appendix B.

It is quite easy now to show that the $(p-1) \times (p-1)$-dimensional matrices A and B so constructed are *Fourier* matrices and that C is the *inverse Fourier* matrix. We see, therefore, that the so-called *fast convolution method* for computing cyclic convolution by taking the product of discrete Fourier transforms followed by inverse discrete Fourier transformation is a special case of the general technique presented here when the field of constants is expanded to include the Nth roots of unity. The interested reader may consult the large body of literature on discrete Fourier transforms over finite fields and their connection to cyclic convolutions; references [3.3, 3.23 - 3.28] are just a representative sample.

3.1.4 Length N Algorithms Over a Field of Constants GF(p) When (N, p) = 1

In this section, we show that the design procedure developed can be employed for cyclic convolution of *any* length N relatively prime to p, i.e., for which $(N, p) = 1$, where the shortened notation denotes the greatest common divisor. This greatly increases the usefulness of the procedure. The following result is necessary; we omit the proof as it is straightforward.

LEMMA 3.2 *If* $(N, p) = 1$, *there exists a positive integer n such that* $N | (p^n - 1)$.

We assume that a cyclic convolution algorithm $C(A\mathbf{x} \bullet B\mathbf{y})$ has been designed for length $p^n - 1$, and we wish to construct an algorithm for length N, $(N, p) = 1$. Let $N' = (p^n - 1)/N$ and define the index subset S_N as

$$S_N = \{i \epsilon S : N' | i\},$$

where S is the index set associated with the length $(p^n - 1)$ cyclic convolution. We are obviously concerned in this section with the case $N \neq p^n - 1$. The following theorem allows us to obtain the algorithm for length N.

THEOREM 3.4 *The cyclic convolution ψ of the length N sequences* $\mathbf{x}$ *and* $\mathbf{y}$ *over* $GF(p^m)$ *can be obtained as*

$$\psi = C'(A'\mathbf{x} \bullet B'\mathbf{y}),$$

where

$$A' = [A_{i_1}'^T, \ A_{i_2}'^T, \cdots]^T, \ B' = [B_{i_1}'^T, \ B_{i_2}'^T, \cdots]^T, \ and$$

$$C' = (N' \bmod p)[C_{i_1}', \ C_{i_2}', \cdots],$$

with $N' = (p^n - 1)/N$ and $i_1, i_2, \ldots \epsilon S_N$. The A_i' and B_i' matrices, $i \epsilon S_N$, are the matrices formed by the first N columns of A_i and B_i, respectively. The C_i' matrices, $i \epsilon S_N$, are the matrices formed by the first N rows of C_i.

PROOF. Define the length $p^n - 1$ vectors $\tilde{\mathbf{x}} = [\mathbf{x}^T \mathbf{x}^T \cdots \mathbf{x}^T]^T$ and $\tilde{\mathbf{y}} = [\mathbf{y}^T \mathbf{y}^T \cdots \mathbf{y}^T]^T$ by repeating $\mathbf{x}$ and $\mathbf{y}$ N' times, respectively. The vector $\tilde{\psi} = \tilde{\mathbf{x}} * \tilde{\mathbf{y}}$ is periodic with period N and

$$\tilde{\psi} = (N' \bmod p)\psi.$$

The algorithm for length $p^n - 1$ is given by

$$\tilde{\psi} = C(A\tilde{\mathbf{x}} \bullet B\tilde{\mathbf{y}}),$$

where

$$A\tilde{\mathbf{x}} = [\tilde{A}_{i_1}^T, \ \tilde{A}_{i_2}^T, \cdots]^T \mathbf{x},$$

with

$$(\tilde{A}_i)_{kj} = \sum_{\ell=0}^{N'-1} (A_i)_{k,j+\ell N}, \quad i \epsilon S.$$

A similar relation holds for the term $B\tilde{\mathbf{y}}$.

We have $\tilde{A}_0 = (N' \bmod p)A_0$. For any nonzero $i \epsilon S$, it is possible from $P3$ to find a $\theta = (p^n - 1)/(p^{\sigma_\theta} - 1)$ such that i is a multiple of θ. From Theorem 3.3, we also know that the rows of A_θ are the MLRS's generated by $P_\theta(u)$. We may then find a $\beta_k \epsilon GF(p^n)$ that satisfies

$$(A_\theta)_{kj} = tr(\beta_k \alpha^{\theta j}), \quad j = 0, 1, \ldots, n-1.$$

From Theorem 3.2, we may obtain the rows of A_i by sampling those of A_θ with period (i/θ); thus,

$$(A_i)_{kj} = tr(\beta_k \alpha^{ij}).$$

Equating the above expressions and using the fact that $tr(\cdot)$ is linear, we obtain

$$(\tilde{A}_i)_{kj} = tr\left(\beta_k \alpha^{ij} \sum_{\ell=0}^{N'-1} \alpha^{i\ell N}\right)$$

which immediately leads to

$$(\tilde{A}_i)_{kj} = \begin{cases} (N' \bmod p)(A_i)_{kj}, & \alpha^{iN} = 1; \\ 0, & \alpha^{iN} \neq 1. \end{cases}$$

A similar expression holds for the relationship of $\tilde{B}_i$ to B_i. Noting that the conditions $i \epsilon S$ and $\alpha^{iN} = 1$ together are equivalent to $i \epsilon S_N$, completes the proof.

EXAMPLE 3. In a practical application of length N cyclic convolution algorithm design from an algorithm of length $(p^n - 1)$, the smallest n that satisfies $N | (p^n - 1)$ should be used. We deal with the field $GF(3^m)$ in this example and first design the larger cyclic convolution from which the smaller will be obtained by application of the preceding results. Let $m = n = 2$; thus, the larger cyclic convolution is of length $3^n - 1 = 8$. We assume the smaller to be of length $N = 4$.

For the length 8 convolution, the cyclotomic sets and orders are $S_0 = \{0\}$, $\sigma_0 = 1$; $S_1 = \{1, 3\}$, $\sigma_1 = 2$; $S_2 = \{2, 6\}$, $\sigma_2 = 2$; $S_4 = \{4\}$, $\sigma_4 = 1$; and $S_5 = \{5, 7\}$, $\sigma_5 = 2$. The index set $S = \{0, 1, 2, 4, 5\}$. Let α denote the primitive element of $GF(3^2)$ satisfying $\alpha^2 + \alpha + 2 = 0$. The divisors t of n are 1 and 2; thus, $\theta = 4$ and 1, respectively. For $\theta = 1$ and 4, the primitive polynomials are $P_1(u) = u^2 + u + 2$ and $P_4(u) = u + 1$, and the (monic) reciprocal polynomials are $\hat{P}_1(u) = u^2 + 2u + 2$ and $\hat{P}_4(u) = u + 1$, respectively. In determining the first $\sigma_1 = 2$ elements of each row of A_1, we use the degree $\sigma_1 - 1 = 1$ polynomial multiplication algorithm of Appendix B. In determining the first $\sigma_1 = 2$ rows of C_1, the following values for $tr(\cdot)$ over $GF(3^2)$ are required:

$$tr(\alpha^i) = \begin{cases} 2, & i = 0,1,3; \\ 1, & i = 4,5,7; \\ 0, & i = 2,6. \end{cases}$$

Since $\sigma_4 = 1$, computation of the first element of A_1 and C_1 is trivial.

The reader may easily verify that the matrices of the length 8 algorithm $C(Ax \bullet By)$ are obtained as $A = B = [\mathbf{A}_0^T, A_1^T, A_2^T, \mathbf{A}_4^T, A_5^T]^T$, $\mathbf{A}_0 = [11 \cdots 1]$, and $C = [\mathbf{C}_0, C_1, C_2, \mathbf{C}_4, C_5]$, $\mathbf{C}_0 = 2\mathbf{A}_0^T$, where

$$A_1 = \begin{bmatrix} 1 & 0 & 1 & 2 & 2 & 0 & 2 & 1 \\ 0 & 1 & 2 & 2 & 0 & 2 & 1 & 1 \\ 1 & 1 & 0 & 1 & 2 & 2 & 0 & 2 \end{bmatrix},$$

$$A_2 = \begin{bmatrix} 1 & 1 & 2 & 2 & 1 & 1 & 2 & 2 \\ 0 & 2 & 0 & 1 & 0 & 2 & 0 & 1 \\ 1 & 0 & 2 & 0 & 1 & 0 & 2 & 0 \end{bmatrix},$$

$$\mathbf{A}_4 = \begin{bmatrix} 1 & 2 & 1 & 2 & 1 & 2 & 1 & 2 \end{bmatrix},$$

$$A_5 = \begin{bmatrix} 1 & 0 & 1 & 1 & 2 & 0 & 2 & 2 \\ 0 & 2 & 2 & 1 & 0 & 1 & 1 & 2 \\ 1 & 2 & 0 & 2 & 2 & 1 & 0 & 1 \end{bmatrix},$$

$$C_1 = \begin{bmatrix} 0 & 1 & 1 & 2 & 0 & 2 & 2 & 1 \\ 2 & 0 & 2 & 2 & 1 & 0 & 1 & 1 \\ 1 & 1 & 2 & 0 & 2 & 2 & 1 & 0 \end{bmatrix}^T,$$

$$C_2 = \begin{bmatrix} 0 & 1 & 0 & 2 & 0 & 1 & 0 & 2 \\ 2 & 2 & 1 & 1 & 2 & 2 & 1 & 1 \\ 1 & 2 & 2 & 1 & 1 & 2 & 2 & 1 \end{bmatrix}^T,$$

$$\mathbf{C}_4 = \begin{bmatrix} 2 & 1 & 2 & 1 & 2 & 1 & 2 & 1 \end{bmatrix}^T,$$

$$C_5 = \begin{bmatrix} 0 & 2 & 1 & 1 & 0 & 1 & 2 & 2 \\ 2 & 0 & 2 & 1 & 1 & 0 & 1 & 2 \\ 1 & 2 & 2 & 0 & 2 & 1 & 1 & 0 \end{bmatrix}^T.$$

This algorithm requires 11 multiplications over $GF(3)$.

Now, to obtain the smaller length 4 algorithm, note that $N' = (p^n - 1)/N = 2$ and $S_N = \{2, 4\}$. Using the above matrices, A_0, A_2, A_4, C_0, C_2, and C_4, we obtain the matrices of the length 4 algorithm $C'(A'\mathbf{x} \bullet B'\mathbf{y})$ from Theorem 3.4 as

$$
A' = B' = \begin{bmatrix} 1 & 1 & 0 & 1 & 1 \\ 1 & 1 & 2 & 0 & 2 \\ 1 & 2 & 0 & 2 & 1 \\ 1 & 2 & 1 & 0 & 2 \end{bmatrix}^T \quad \text{and} \quad C' = 2 \begin{bmatrix} 2 & 0 & 2 & 1 & 2 \\ 2 & 1 & 2 & 2 & 1 \\ 2 & 0 & 1 & 2 & 2 \\ 2 & 2 & 1 & 1 & 1 \end{bmatrix}.
$$

This algorithm requires 5 multiplications over $GF(3)$.

3.1.5   Multiplicative Complexity

Inherent in the design of bilinear algorithms for cyclic convolution described in the preceding sections is the availability of algorithms for multiplying two $\sigma_i - 1$ degree polynomials for all values of $i \epsilon S_N$ (S_N is the index subset of S defined in Section 3.1.4). We also note that other properties, e.g., symmetry of calculation, as possessed by those algorithms of Appendix B, can also greatly simplify the design procedure. Counting only the multiplications involved in the computation of $\mathbf{m}$ in $\psi = C(A\mathbf{x} \bullet B\mathbf{y}) = C\mathbf{m}$, we may easily write the multiplicative complexity of the length N cyclic convolution algorithm, $M(N)$ (we avoid the previous use of n here, so as not to confuse with the exponent of p in $p^n - 1$), as

$$
M(N) = 1 + \sum_{i \epsilon S_N} M_{\sigma_i}, \tag{3.8}
$$

where M_{σ_i} denotes the multiplicative complexity of degree $(\sigma_i - 1)$ polynomial multiplication. Tables 3.1 - 3.4 from [3.17] list the values of $M(N)$ and the ratio $M(N)/N$ for various length N cyclic convolution algorithms designed as described here over $GF(2)$ and $GF(3)$. Tables 3.3 and 3.4 refer to the more general *factor length* procedure of Section 3.1.4, e.g., the length $N = 5$ algorithm of Table 3.3 is based on an algorithm of length $p^n - 1$, where $p = 2$ and $n = 4$.

The bilinear algorithm design procedure remains valid if p is replaced by p^m; this increases the order of the field of constants and thereby reduces the number of multiplications required [3.29]. Table 3.5, from [3.17], shows the multiplicative complexity over $GF(4)$ for

Table 3.1: Complexity of Cyclic Convolution Algorithm Over a Field of Constants $GF(2)$.

Length N	No. of Mults. $M(N)$	$M(N)/N$
3	4	1.3333
7	13	1.8571
15	31	2.0667
63	178	2.8254
255	841	3.2980
511	2029	3.9707
4095	18295	4.4676

Table 3.2: Complexity of Cyclic Convolution Algorithm Over a Field of Constants $GF(3)$.

Length N	No. of Mults. $M(N)$	$M(N)/N$
2	2	1
8	11	1.375
26	50	1.9231
80	173	2.1625
728	2147	2.9492

Table 3.3: Complexity of Cyclic Convolution of Some Factor Lengths Over a Field of Constants $GF(2)$.

Length N	No. of Mults. $M(N)$	$M(N)/N$
5	10	2
9	22	2.4444
13	55	4.2308
17	55	3.2353
21	52	2.4762
35	130	3.7143
45	157	3.4889
51	166	3.2549
65	280	4.3077
73	289	3.9589
85	280	3.2941
117	508	4.3419
315	1285	4.0794

Table 3.4: Complexity of Cyclic Convolution of Some Factor Lengths Over a Field of Constants $GF(3)$.

Length N	No. of Mults.$M(N)$	$M(N)/N$
4	5	1.25
5	10	2.0
7	19	2.7143
10	20	2.0
13	25	1.9231
14	38	2.7143
16	29	1.8125
20	41	2.05
28	77	2.75
40	83	2.075
52	125	2.4039
56	155	2.7679
91	259	2.8462
104	275	2.6442
182	518	2.8462
362	1061	2.9148

Table 3.5: Complexity of Cyclic Convolution Algorithm Over a Field of Constants $GF(4)$.

Length N	No. of Mults. $M(N)$	$M(N)/N$
15	21	1.4
63	123	1.95234
255	561	2.2
4095	12201	2.9795

several cyclic convolution lengths also present in Table 3.1. Just a factor of two increase in field size is seen to substantially decrease (by approximately 30%) the multiplicative complexity; however, computation of the linear forms associated with the matrices A, B, and C whose elements are now over the enlarged field is somewhat more complex. Current research into the specialized hardware and procedures of *Galois field computers* will permit efficient computation over large finite fields.

Some comparison with results obtained in the literature is in order. We first discuss the *theoretical minimum* number of multiplications. The generating polynomials associated with the indeterminate sequences x_i and y_i, $i = 0, 1, \ldots, N - 1$, are given by

$$X(u) = \sum_{i=0}^{N-1} x_i u^i \text{ and } Y(u) = \sum_{i=0}^{N-1} y_i u^i. \tag{3.9}$$

The cyclic convolution $\psi = \mathbf{x} * \mathbf{y}$ may then be obtained as the coefficients of the polynomial

$$\Psi(u) = X(u)Y(u) \bmod P(u), \tag{3.10}$$

where $P(u) = u^N - 1$. If $P(u) = \prod_{i=1}^{L} P_i(u)$, where $P_i(u)$ is an irreducible polynomial, then the length N problem (3.10) may be broken up into L smaller problems of the form

$$\Psi_i(u) \equiv X_i(u)Y_i(u) \bmod P_i(u), \quad i = 1, 2, \ldots, L, \tag{3.11}$$

where $X_i(u) \equiv X(u) \bmod P_i(u)$, $Y_i(u) \equiv Y(u) \bmod P_i(u)$, and from which (unique) reconstruction of $\Psi(u)$ is possible using the Chinese Remainder Theorem, i.e.,

$$\Psi(u) = \left[\sum_{i=1}^{L} S_i(u)\Psi_i(u)\right] \bmod P(u). \tag{3.12}$$

The polynomials $S_i(u)$, $i = 1, 2, \ldots, L$, satisfy the congruences

$$S_i(u) \equiv \begin{cases} 0 & \bmod P_j(u),\ j = 1, 2, \ldots, L, j \neq i; \\ 1 & \bmod P_i(u), \end{cases} \tag{3.13}$$

and, thus, have the form

$$S_i(u) = R_i(u) \prod_{\substack{j=1 \\ j \neq i}}^{L} P_j(u), \tag{3.14}$$

where

$$R_i(u) \prod_{\substack{j=1 \\ j \neq i}}^{L} P_j(u) \equiv 1 \bmod P_i(u). \tag{3.15}$$

The coefficients of each polynomial $S_i(u)$ lie in the same field $\mathcal{F}$ as the coefficients of the $P_i(u)$. If $\mathcal{F} = GF(p)$, the above residue reduction operations are performed over $GF(p)$; furthermore, the factorization of $P(u) = u^N - 1$ into L irreducible factors must be carried out over $GF(p)$.

In [3.29], it is shown that the *minimum* number of multiplications required to compute (3.10) is $\mu = 2 \deg[P(u)] - L = 2N - L$; moreover, every minimal algorithm for computing (3.10) is *disjoint* (computes smaller systems of bilinear forms on disjoint sets of indeterminates) and is based on the CRT. Note that the choice $P(u) = u^N - 1$ is specialized to the present problem of cyclic convolution, but the above remarks are not.

It is well known that the polynomial $u^{p^n} - u$ equals the product of all distinct, monic, irreducible polynomials over $GF(p)$ whose degrees divide n. A problem closely connected to this factorization (and to cyclic codes) is the factorization of $P(u) = u^N - 1$ over $GF(p)$. If $(N, p) = 1$, $u^N - 1$ factors into the product of distinct minimal polynomials of the elements α^i, $i = 0, 1, \ldots, N - 1$; thus, the zeroes of the minimal polynomial of α^i are elements of the finite set $\{\alpha^{ip}, \alpha^{ip^2}, \ldots\}$, where the exponents are understood to be mod N. We can partition the elements α^i into sets, each set corresponding to an irreducible factor of $u^N - 1$. Equivalently partitioning the integers $[0, N-1]$ in this manner, results in each element of S_N as corresponding to an irreducible factor of $u^N - 1$; thus, $L = |S_N| + 1$. This leads us to the theoretical minimum number of multiplications for a length N cyclic convolution algorithm over $GF(p)$ as being $\mu = 2N - |S_N| - 1$. Such minimal algorithms are generally difficult to derive for large N and may also require a large number of additions.

As an example which also illustrates a convenient method for finding the factors of $u^N - 1$, consider $N = 15$. In Example 1, we found that the index set $S = \{0, 1, 3, 5, 7\} = S_N \cup \{0\}$. One way of finding the factors of $u^{15} - 1$ is to first assume that $\mathcal{F} = \mathcal{C}$, the field of complex numbers, and factor $u^{15} - 1$ using *cyclotomic polynomials*. We obtain

$$u^N - 1 = \prod_{\substack{1 \le d \le N \\ d|N}} C_d(u) = C_1(u)\, C_3(u)\, C_5(u)\, C_{15}(u), \quad N = 15,$$

where

$$C_1(u) = u - 1, \; C_3(u) = u^2 + u + 1, \; C_5(u) = u^4 + u^3 + u^2 + u + 1,$$
$$\text{and } C_{15}(u) = u^8 - u^7 + u^5 - u^4 + u^3 - u + 1.$$

We now determine if the $C_d(u)$ can be further factored over $GF(p)$; let $p = 2$. The polynomials $C_d(u)$, $d = 1, 3, 5$, are easily seen to be irreducible over $GF(2)$. Since $C_{15}(u)$ is a polynomial of degree 8 and $15|(2^4 - 1)$, $C_{15}(u)$ factors into two polynomials of degree 4, both of which are irreducible over $GF(2)$, i.e.,

$$C_{15}(u) = (u^4 + u^3 + 1)\,(u^4 + u + 1).$$

These results imply that $\mu = 25$ multiplications is the theoretical minimum number for $N = 15$ over $GF(2)$. With the aid of the cyclotomic polynomial factorization over $\mathcal{F} = \mathcal{C}$, the interested reader may examine the complexity associated with other lengths.

In terms of other available cyclic convolution algorithms, the one developed here either has lower or very competitive multiplicative complexity, e.g., in [3.30] an algorithm for length $N = 15$ over the field of constants $GF(2)$ requires 40 multiplications. Reducing every constant modulo 2 in the rational field algorithms of [3.1], a cyclic convolution over $GF(2)$ of length $N = 63$ requires 418 multiplications (since in this case, $S = S_N \cup \{0\}$ and $|S_N| = 13$, the theoretical minimum, $\mu = 113$ multiplications). It is true, however, that because of the transformation of the constants, the algorithms adapted to finite fields might be computationally less complex than the original algorithms over fields such as the rationals, but it is generally difficult to identify the redundant multiplications.

3.1.6   Additive Complexity

The *structure* of the procedure developed here may also be used to reduce the additive complexity in calculating the linear forms $A\mathbf{x}$, $B\mathbf{y}$, and $C\mathbf{m}$. This, most definitely, sets it apart from most known procedures. Some observations on this are

1. From Theorem 3.3, for every $i \epsilon S_N$, the rows of A_i are MLRS's over the field of constants $GF(p)$ based on the same primitive polynomial of degree n, where n is the smallest integer such that $N | (p^n - 1)$. Since only n MLRS's generated by a primitive polynomial of degree n may be linearly independent over $GF(p)$, only n rows of A_i, $i \epsilon S_N$, are linearly independent. It follows that to compute $A_i\mathbf{x}$, it is only necessary to compute n elements and obtain the remaining components from these. The balance property of MLRS's [3.22] can be used to reduce the complexity of the initial n computations. The same observation is true for the computation $B\mathbf{y}$.

2. If $i \epsilon S_N$, $i | N$, then all the rows of A_i are periodic with period N/i, by Theorem 3.2. To form $A_i\mathbf{x}$, we may first add every (N/i)th component of $\mathbf{x}$ and then multiply the reduced size vector with the nonperiodic portion of A_i. The same observation is true for the computation $B\mathbf{y}$.

3. For any $i \epsilon S_N$, the columns of C are MLRS's based on the same primitive polynomial. If n is the smallest integer such that $N | (p^n - 1)$, then any row of C_i can be expressed as a linear sum over

$GF(p)$ of at most n immediately previous rows. The linear relation is the difference equation based on the minimal polynomial of α^{-i}. A strategy can easily be devised to exploit this structure.

4. If $i\epsilon S_N$ and $i|N$, then the rows of C_i are periodic with period N/i. This implies that $C_i\mathbf{m}_i$ ($\mathbf{m}_i$ is that portion of $\mathbf{m}$ which multiplies C_i) is also periodic with the same period. Once again, a strategy can easily be devised to exploit this structure.

Application of the above methods provides a systematic way of reducing additive complexity. For example, the length $N = 5$ algorithm of Table 3.3 requires 10 multiplications and only 27 additions over $GF(2)$. This additive complexity is easy to obtain using the above suggestions. By way of comparison, adapting commonly used algorithms over the rationals results in a higher additive complexity of 31 additions (with 10 multiplications). Optimizing the algorithm so that the number of multiplications is the theoretical minimum, $\mu = 2N - |S_N| - 1 = 8$ ($|S_N| = 2$), can more than double the additive complexity.

3.2 Aperiodic Convolution

In Section 3.1, our concern was with the systematic design of a bilinear algorithm for length N cyclic convolution over a field of constants $GF(p)$ when $(N, p) = 1$. Comparison was made against CRT-based algorithms having the theoretical minimum number of multiplications and several suggestions were made for further reducing additive complexity. In general, a polynomial product $Z(u)Y(u) \bmod P(u)$ cannot be performed in fewer than $\mu = 2 \deg [P(u)] - L$ multiplications, where L is the number of relatively prime polynomial factors of $P(u)$ over the field of constants $GF(p)$. Our plan in this section is to first briefly investigate the algorithm construction leading to the theoretical minimum number of multiplications for aperiodic convolution; study the applicability of the algorithms of Section 3.1 to aperiodic convolution by embedding, and compare this approach to multidimensional methods; and finally, present a *heuristic* approach to efficient aperiodic convolution design over the field of constants $GF(p)$.

3.2.1   The Toom-Cook Algorithm

Consider the indeterminate sequences, $\{z_0, z_1, \ldots, z_{k-1}\}$ and $\{y_0, y_1, \ldots, y_{d-1}\}$; let the respective generating polynomials be

$$Z(u) = \sum_{i=0}^{k-1} z_i u^i \text{ and } Y(u) = \sum_{i=0}^{d-1} y_i u^i.$$

We are interested in the product $\Phi(u) = Z(u)Y(u)$. The $k + d - 1$ coefficients of $\Phi(u)$ correspond to a system of bilinear forms $\phi = Z\mathbf{y}$; we have seen an example of such a system in Section 2.4. In that context, it was formulated as the A-dual of another system of bilinear forms associated with the computation of FIR filter outputs. One way of computing ϕ using $k+d-1$ multiplications first requires the selection of $k+d-1$ *distinct* constants, $a_0, a_1, \ldots, a_{k+d-2}$, from the relevant field, $\mathcal{F}$. It is clear that

$$\Phi(u) = Z(u)Y(u) = Z(u)Y(u) \bmod P(u),$$

where $\deg[P(u)] > \deg[\Phi(u)] = k+d-2$. Choose $P(u)$ as the product of distinct, degree-one factors formed using the above constants, i.e.,

$$\Phi(u) = Z(u)Y(u) \bmod \left[\prod_{i=0}^{k+d-2} (u - a_i) \right]. \tag{3.16}$$

The CRT applies easily to this problem. We first compute the subproblems, $Z(u)Y(u) \bmod (u - a_i) = Z(a_i)Y(a_i)$, $i = 0, 1, \ldots, k+d-2$, then reconstruct the solution using only multiplications by elements of $\mathcal{F}$ (which are not counted in the multiplicative complexity). The multiplicative complexity is then exactly $\mu = k + d - 1$, the effort required in the computation of $Z(a_i)Y(a_i)$, $i = 0, 1, \ldots, k + d - 2$. This is the theoretical minimum, as shown in [3.31]; moreover, every minimal algorithm has the above form (to within products of $Z(a_i)$, $Y(a_i)$ by field constants) or uses a slightly different approach. This modified approach will prove to be of interest in the sequel. Assume now that we select $k + d - 2$ distinct constants, $b_1, b_2, \ldots, b_{k+d-2}$, from $\mathcal{F}$. We may then write (3.16) as

$$\Phi(u) = Z(u)Y(u) \bmod \left[\prod_{i=1}^{k+d-2} (u - b_i) \right] + z_{k-1}y_{d-1} \prod_{i=1}^{k+d-2} (u - b_i)$$

$$= \Phi'(u) + z_{k-1} y_{d-1} \prod_{i=1}^{k+d-2} (u - b_i). \tag{3.17}$$

The CRT requires $k + d - 2$ multiplications to compute $\Phi'(u)$, and the computation $z_{k-1} y_{d-1}$ requires one multiplication, for a total once more of $\mu = k + d - 1$.

The process of (3.17) is similar to that of intentionally allowing polynomial coefficient *wraparound*. Conventional methods, plus the wraparound concept (which would, in general, first compute $Z(u)Y(u)$ mod $P(u)$, where $\deg[P(u)] = D < k + d - 1$), can sometimes lead to algorithms having very close to minimum multiplicative complexity. We will see examples of this in Section 3.2.3. The algorithm synthesis method of (3.16) is referred to as the Toom-Cook algorithm [3.32-3.33] and is closely tied to the fact that an $(N - 1)$th order polynomial can be exactly interpolated through N points using Lagrange's interpolation formula. There are practical problems with this approach: first, we must have $|\mathcal{F}| > k + d - 2$, i.e., the field of constants must be large enough to ensure that N distinct constants can be found, and second, the constants must be of simple form; otherwise, the additive complexity may be large.

3.2.2 Embedding of Aperiodic Convolution: Higher-Dimensional Approaches

The most commonly encountered form of convolution is probably of the *casual* aperiodic type. We have come across this form of convolution before in regard to the derivation of the A-dual system in Section 2.4 , where a casual aperiodic convolution of length $N = k + d - 1$ was obtained from two indeterminate sequences of lengths k and d. To apply *cyclic* convolution to this problem, we can extend both indeterminate sequences to length N by padding zeroes and then cyclically convolve these extended sequences. For example, this embedding approach can be taken to handle the A-dual computation in Example 2 of Chapter 2 for which $k = d = 4$, $N = 7$. Of course, if $k = d = 2^r$, as is the case in the example referred to, we can employ a cyclic convolution of length $2^{r+1} - 1$ to compute the aperiodic convolution. Doing this requires a cyclic convolution of length 7, which has 13 multiplications over $GF(2)$, from Table 3.1.

Another approach is to apply multidimensional techniques [3.34].

Consider the causal aperiodic convolution of the indeterminate sequences, $\{z_0, z_1, \ldots, z_{k-1}\}$ and $\{y_0, y_1, \ldots, y_{d-1}\}$, each extended to a length N; it can be written with the usual assumptions as

$$\phi_t = \sum_{i=0}^{N-1} z_{t-i} y_i, \quad t = 0, 1, \ldots, 2N - 2.$$

We assume that the integer N factors into a product of integers as $N = N_1 N_2$ and we represent the range of the indices t, i as $t = l + mN_1$, $i = p + qN_2$, respectively. This allows us to write the aperiodic convolution in the following manner:

$$\tilde{\phi}_{l,m} = \sum_{q=0}^{N_2-1} \sum_{p=0}^{N_1-1} \tilde{z}_{l-p,m-q} \, \tilde{y}_{p,q}, \quad l = 0, 1, \ldots, 2N_1 - 2;$$
$$m = 0, 1, \ldots, 2N_2 - 2. \quad (3.18)$$

The coefficients, $\tilde{z}_{p,q} = z_{p+qN_1}$ and $\tilde{y}_{p,q} = y_{p+qN_1}$, $p = 0, 1, \ldots, N_1 - 1$ and $q = 0, 1, \ldots, N_2 - 1$, can be thought of as elements of $(N_1 \times N_2)$-dimensional arrays. The collection of coefficients $\tilde{\phi}_{l,m}$ comprises a $(2N_1 - 1) \times (2N_2 - 1)$-dimensional array. It is not difficult to show that the desired aperiodic convolution coefficients can be reconstructed using additions from

$$\phi_t = \tilde{\phi}_{l,m} + \tilde{\phi}_{l+N_1,m-1}, \quad t = l + mN_1;$$
$$l = 0, 1, \ldots, 2N_1 - 2;$$
$$m = 0, 1, \ldots, 2N_2 - 2. \quad (3.19)$$

It is clear from the above expressions that the procedure is a form of the popular *overlap-add* method used for block processing.

If $N = 2^r$, the problem may be formulated as an r-dimensional convolution, similar to the *select-save* method, also used for block processing. Both the overlap-add and select-save methods fall in the category of what are known as sectioning techniques. We now consider an example of the above approach.

EXAMPLE 4. Let $N = N_1 N_2 = 2^r$, $r = 2$, and $N_1 = N_2$; the aperiodic convolution has the form

$$\phi = \begin{bmatrix} z_0 & 0 & 0 & 0 \\ z_1 & z_0 & 0 & 0 \\ z_2 & z_1 & z_0 & 0 \\ z_3 & z_2 & z_1 & z_0 \\ 0 & z_3 & z_2 & z_1 \\ 0 & 0 & z_3 & z_2 \\ 0 & 0 & 0 & z_3 \end{bmatrix} \begin{bmatrix} y_0 \\ y_1 \\ y_2 \\ y_3 \end{bmatrix} ;$$

this is the $k = 4$, $d = 4$ example of Section 2.4. The $\tilde{z}_{p,q}$ and $\tilde{y}_{p,q}$ form arrays with N_2 columns consisting of blocks of length N_1 from the indeterminate sequences, i.e.,

$$\tilde{Z} = \begin{bmatrix} \tilde{z}_{0,0} & \tilde{z}_{0,1} \\ \tilde{z}_{1,0} & \tilde{z}_{1,1} \end{bmatrix} = \begin{bmatrix} z_0 & z_2 \\ z_1 & z_3 \end{bmatrix}, \quad \tilde{Y} = \begin{bmatrix} \tilde{y}_{0,0} & \tilde{y}_{0,1} \\ \tilde{y}_{1,0} & \tilde{y}_{1,1} \end{bmatrix} = \begin{bmatrix} y_0 & y_2 \\ y_1 & y_3 \end{bmatrix}.$$

From (3.18), we compute the $(2N_1 - 1) \times (2N_2 - 1)$-dimensional matrix $\tilde{\Phi}$ as

$$\tilde{\Phi} = \begin{bmatrix} \tilde{\phi}_1 | \tilde{\phi}_2 | \tilde{\phi}_3 \end{bmatrix},$$

where

$$\tilde{\phi}_1 = \begin{bmatrix} \tilde{z}_{0,0}\tilde{y}_{0,0} \\ \tilde{z}_{1,0}\tilde{y}_{0,0} + \tilde{z}_{0,0}\tilde{y}_{1,0} \\ \tilde{z}_{1,0}\tilde{y}_{1,0} \end{bmatrix} = \begin{bmatrix} z_0 y_0 \\ z_1 y_0 + z_0 y_1 \\ z_1 y_1 \end{bmatrix},$$

$$\tilde{\phi}_2 = \begin{bmatrix} \tilde{z}_{0,1}\tilde{y}_{0,0} + \tilde{z}_{0,0}\tilde{y}_{0,1} \\ \tilde{z}_{1,1}\tilde{y}_{0,0} + \tilde{z}_{0,1}\tilde{y}_{1,0} + \tilde{z}_{1,0}\tilde{y}_{0,1} + \tilde{z}_{0,0}\tilde{y}_{1,1} \\ \tilde{z}_{1,1}\tilde{y}_{1,0} + \tilde{z}_{1,0}\tilde{y}_{1,1} \end{bmatrix}$$

$$= \begin{bmatrix} z_2 y_0 + z_0 y_2 \\ z_3 y_0 + z_2 y_1 + z_1 y_2 + z_0 y_3 \\ z_3 y_1 + z_1 y_3 \end{bmatrix},$$

$$\tilde{\phi}_3 = \begin{bmatrix} \tilde{z}_{0,1}\tilde{y}_{0,1} \\ \tilde{z}_{1,1}\tilde{y}_{0,1} + \tilde{z}_{0,1}\tilde{y}_{1,1} \\ \tilde{z}_{1,1}\tilde{y}_{1,1} \end{bmatrix} = \begin{bmatrix} z_2 y_2 \\ z_3 y_2 + z_2 y_3 \\ z_3 y_3 \end{bmatrix}.$$

Table 3.6: Complexity of Aperiodic Convolution Over a Field of Constants $GF(2)$ for Lengths of the Form 2^r.

	No. of Mults.	
Length (2^r)	Embed	Multidim.
2	4	3
4	13	9
8	31	27
32	178	243
128	841	2187
256	2029	6561

The reader may then easily verify that (3.19) applied to the elements $\tilde{\phi}_{l,m}$ of $\tilde{\Phi}$ provides the correct $2N - 1 = 7$ aperiodic convolution coefficients (note that ϕ_2 and ϕ_4 will appear twice due to the fact that the mapping $t \to (l, m)$ is not one-to-one). Along each dimension, it is important to efficiently perform an aperiodic convolution of two length 2 sequences; this can be carried out using 3 multiplications and 4 additions. The total multiplicative complexity for N of the form 2^r is, therefore, 3^r and 9 for this example.

Table 3.6 provides a comparison, for various aperiodic convolution lengths, of the number of multiplications required over the field of constants $GF(2)$ by the method which embeds aperiodic convolution in cyclic convolution (computed as in Section 3.1) and the multidimensional sectioning approach. The procedure described in [3.34] can also be used for aperiodic convolution where the length is of the form $2^r 3^s$. Table 3.7 presents a complexity comparison for a short list of representative lengths interleaving those of Table 3.6.

Finally, it is important to realize that many algorithms do not take sufficient advantage of a certain freedom that exists in selecting the field of constants over which the computations are performed. Table

Table 3.7: Complexity of Aperiodic Convolution Over a Field of Constants $GF(2)$ for Lengths of the Form $2^r 3^s$.

| | No. of Mults. | |
Length ($2^r 3^s$)	Embed	Multidim.
3	13	6
6	31	18
18	156	108
48	403	486
64	841	729
96	841	1458
192	2029	4374

3.8 illustrates the savings possible when the aperiodic convolution is computed over the field of constants $GF(3)$ as opposed to $GF(2)$.

As we previously mentioned, the index mappings encountered in the multidimensional approach are not one-to-one; this leads to a certain computational inefficiency on one hand and a certain flexibility on the other, since the factors of N need not be mutually prime (in [3.34], this permitted the use of two-dimensional Fermat number transformations). The efficiency of these algorithms is, however, still quite respectable, particularly for small-to-moderate sizes.

If the problem at hand is, in fact, the aperiodic convolution of length k and length d sequences, a clear inefficiency arises in first extending these both to length N. We can, at least, avoid this if k and d have a common factor, i.e., if $k = k_1 n_1$, $d = d_1 n_1$. This approach, in one sense, generalizes the multidimensional technique described above. Consider the two-dimensional case, i.e., k_1 and d_1, themselves, have no common factor; the generating polynomials $Z(u)$ and $Y(u)$ of the respective indeterminate sequences can be written for $v = u^{n_1}$ as

$$Z(u) = Z(u,v) = \sum_{p=0}^{n_1-1} Z_p(v)u^p,$$

Table 3.8: Complexity of Aperiodic Convolution Over Fields of Constants $GF(2)$ and $GF(3)$.

		No. of Mults.	
Length	Embed, $GF(2)$	Embed, $GF(3)$	Multidim.
3	13	11	6
9	52	50	36
12	93	50	54
24	178	173	162
27	178	173	216
36	403	173	324
288	6087	2147	8748

and

$$Y(u) = Y(u, v) = \sum_{q=0}^{n_1-1} Y_q(v) u^q,$$

where

$$Z_p(v) = \sum_{m=0}^{k_1-1} z_{p+mn_1} v^m,$$

and

$$Y_q(v) = \sum_{l=0}^{d_1-1} y_{q+ln_1} v^l.$$

The data index mappings used are obviously, $i \rightarrow p + mn_1$, $p = 0, 1, \ldots, n_1 - 1$; $m = 0, 1, \ldots, k_1 - 1$, for the z-sequence and $i \rightarrow q + ln_1$, $q = 0, 1, \ldots, n_1 - 1$; $l = 0, 1, \ldots, d_1 - 1$, for the y-sequence.

The product of $Z(u, v)$ and $Y(u, v)$ produces the two-dimensional polynomial

$$\Phi(u, v) = \sum_{q=0}^{n_1-1} \sum_{p=0}^{n_1-1} Z_p(v) \, Y_q(v) \, u^{p+q}. \tag{3.20}$$

We may then obtain the generating polynomial of the desired aperiodic convolution sequence as $\Phi(u) = \Phi(u, u^{n_1})$. The computation (3.20) of $\Phi(u, v)$ involves the product of two $(n_1 - 1)$-degree polynomials, in which each multiplication is the product of a polynomial of degree $(k_1 - 1)$ with one of degree $(d_1 - 1)$.

The total multiplicative complexity is then $M_{n_1,n_1} M_{k_1,d_1}$, where $M_{(\cdot,\cdot)}$ denotes the multiplicative complexity for sequences having the indicated lengths. In our previous example, $k = d = 4$ and $n_1 = 2$; using the correct algorithm of Appendix B, we again have 9 multiplications. The extension of this procedure to higher dimensions is straightforward.

3.2.3 Aperiodic Convolution with Wraparound

We conclude this chapter with a somewhat heuristic approach to aperiodic convolution algorithm design based, in part, on the minimal algorithm exposed in the Toom-Cook method of Section 3.2.1.

To motivate this section, consider, once again, the aperiodic convolution of two length N sequences, as presented in Section 3.2.2. The problem of computing $\Phi(u) = Z(u)Y(u)$ is equivalent to computing $\Phi(u) = Z(u)Y(u) \bmod P(u)$, where $P(u)$ could be selected as $u^D - 1$, as long as $D \geq 2N - 1$. It may be beneficial, however, to choose $D < 2N - 1$ if the desired aperiodic convolution coefficients can be recovered with a few computations. The next result, from [3.19], formalizes this idea.

LEMMA 3.3 *The polynomial product for degree $N - 1$ polynomials, $\Phi(u) = Z(u)Y(u) \bmod P(u)$, where $\deg [P(u)] = D = 2N - 1$, can be evaluated by computing the $D - s$ coefficients of*

$$\Phi_I(u) \equiv Z(u)Y(u) \bmod (u^{D-s} - 1),$$

and the first and last $s/2$ coefficients of $\Phi(u)$, s even, or the first (last) $(s - 1)/2$ and last (first) $(s + 1)/2$ coefficients of $\Phi(u)$, s odd. The coefficients of $\Phi_I(u)$ are related to those of $\Phi(u)$ by the wraparound expression

$$\phi_{It} = \begin{cases} \phi_t + \phi_{t+D-s}, & t \leq s - 1; \\ \phi_t, & s \leq t < D - s. \end{cases}$$

We note that the s coefficients of

$$\hat{\Phi}(u) \equiv \hat{Z}(u)\hat{Y}(u) \bmod u^s,$$

where the reciprocal polynomials

$$\hat{Z}(u) = u^{N-1} Z(1/u) \text{ and } \hat{Y}(u) = u^{N-1}Y(1/u),$$

represent the values necessary to reconstruct $\Phi(u)$ from $\Phi_I(u)$.

The theoretical minimum number of multiplications obtained when the above procedure is used is seen to be $\mu_s = 2 \deg [P_I(u)] - L_I + M_s$, where $P_I(u) = u^{D-s} - 1$, L_I is the number of irreducible factors over $\mathcal{F}$ of $P_I(u)$, and M_s denotes the number of multiplications required to compute the first and last approximately $s/2$ coefficients. For s even, $M_s = ((s+1)/2)^2 - \frac{1}{4}$ and for s odd, $M_s = ((s+1)/2)^2$; the number of additions also increases quadratically with s. This leads to the minimum multiplicative complexity, $\mu_s = 2(D - s) - L_I + M_s$. It is easy to see that the procedure provides favorable results when an integer s can be found such that $M_s < L_I + 2s - L$, where L is the number of irreducible factors over $\mathcal{F}$ of $P(u)$. Finding an optimum s may not be simple; in practice, we only examine the first few values of s starting at $s = 1$.

Table 3.9 shows the two theoretical minima, $\mu = 2D - L$ and μ_s (the latter for the best value of s), for short-length aperiodic convolution over the rationals. The method is actually quite simple and efficient for short lengths; furthermore, the additive complexity does not rise appreciably due to the fact that the tails of any aperiodic convolution (those points affected by this method) involve the smallest relative number of multiplications and additions. The approach can be extended in various ways to longer lengths, as described in [3.19].

EXAMPLE 5. Consider the aperiodic convolution $\phi = z \odot y$ of length $N = 5$ sequences over the rationals. Let $P(u) = u^D - 1$, $D = 9$, which has $L = 3$ cyclotomic polynomial factors. Trying several values of s, we conclude that $s = 3$ is the best choice; thus, $P_I(u) = u^{D-s} - 1 = u^6 - 1$, which has $L_I = 4$. A length 6 bilinear cyclic convolution algorithm can be employed, which requires 8 multiplications, and produces $\Phi_I(u)$. The $s = 3$ wraparound coefficients

Table 3.9: Theoretical Minimum Complexities for Aperiodic Convolution.

	No. of Mults.	
Length	μ	μ_s
2	4	3
3	8	6
4	12	9
5	15	12
6	18	16
7	24	19
8	26	22
9	30	27
10	34	31

of $\Phi_I(u)$ can then be separated by computing $\hat{\Phi}(u)$ which requires 4 multiplications (note $M_s < L_I + 2s - L$). The total multiplicative complexity is 12 multiplications; the additive complexity is 36 additions. For this length, we have, in fact, achieved the theoretical minimum, μ_s.

We now concentrate on applying the preceding method over $GF(p)$ hand-in-hand with the CRT. The aperiodic convolution of a length k with a length d sequence is desired; let $N = k + d - 1$. We take the CRT-based approach of Section 3.2.1, but do *not* assume $P(u)$ factors into degree-one terms; in addition, we do *not* assume $P(u) = u^N - 1$. The problem is to compute

$$\Phi(u) = Z(u)Y(u) \bmod P(u),$$

where deg $[P(u)] = D \geq N$ and $P(u) = \prod_{i=1}^{L} P_i(u)$, with the factors $P_i(u), P_j(u), i, j = 1, 2, \ldots, L; i \neq j$, relatively prime. Denote the multiplicative complexity of the ith subproblem, $\Phi_i(u) \equiv Z_i(u)Y_i(u) \bmod P_i(u)$, by M_i (this is the number of multiplications required for deg $[P_i(u)] - 1$ polynomial multiplication). Then the total multiplicative complexity is clearly $M_N = \sum_{i=1}^{L} M_i$. In addition to choosing an appropriate wraparound, s, for a given N, the lowest M_N will occur if each of the $P_i(u)$ has as low a degree as possible (recall that $\mu = 2 \deg [P(u)] - L$).

Appendix C lists a number of monic polynomials over $GF(2)$ and $GF(3)$ up to degree 5 and degree 3, respectively. For each degree, those polynomials are listed which are not products of two *distinct* polynomials of lower degree. Note that, due to the manner in which we have constructed the list of polynomials for a given degree l, the number of monic *irreducible* polynomials of degree l over $GF(p)$, $I_p(l)$, provides a lower bound on the number of entries. The number $I_p(l)$ itself is bounded as $p^l[1-p^{-(l/2)+1}] / l < I_p(l) \leq (p^l - p) / l$, a fact which follows directly from the representation of $I_p(l)$ in terms of the Möbius function [3.21]. For each degree, the monic irreducible polynomials are identified as those contained within brackets. These polynomials are enumerated over $GF(2)$ and $GF(3)$ in [3.35] and [3.36], respectively.

We select the greatest possible number of factors from this list in order to form a product polynomial $P_I(u)$ of degree $D - s$, $D = N = k + d - 1$, where s is jointly selected so as to minimize the multiplicative complexity. The list of possible $P_i(u)$ depends on $\mathcal{F}$ and how we handle the computation for each $P_i(u)$ can depend on the specific form of that factor. In general, however, we simply use the bilinear small degree polynomial multiplication algorithms of Appendix B for each subproblem. The s required values are then computed to recover the first and last approximately $s/2$ coefficients of $\Phi(u)$. We provide an example from which some interesting conclusions can be drawn.

EXAMPLE 6.   Consider the aperiodic convolution over $GF(2)$ of two sequences of length $k = d = 4$; thus, $N = k + d - 1 = 7$. We choose $D = N$ and $s = 1$. The factors of $P_I(u)$ are selected from Appendix C starting from the lowest degree and making certain not to include a factor divided by a previously selected one. An appropriate choice is $P_1(u) = u + 1$, $P_2(u) = u^2 + u + 1$, and $P_3(u) = u^3$, for which $\deg [P_I(u)] = N - 1 = 6$.

We perform the decomposition as in (3.11) with respect to the factors $P_i(u)$, $i = 1, 2, 3$, of $P_I(u)$; these require the bilinear algorithms of degrees zero, one, and two, respectively. In order, the results are

$$\Phi_{I1}(u) \equiv Z_1(u)Y_1(u) \bmod P_1(u)$$
$$= m_0,$$

where

$$m_0 = (z_0 + z_1 + z_2 + z_3)(y_0 + y_1 + y_2 + y_3);$$

$$\Phi_{I2}(u) \equiv Z_2(u)Y_2(u) \bmod P_2(u)$$
$$= (m_1 + m_3)u + (m_1 + m_2),$$

where

$$\begin{aligned}
m_1 &= (z_0 + z_2 + z_3)(y_0 + y_2 + y_3), \\
m_2 &= (z_1 + z_2)(y_1 + y_2), \\
m_3 &= (z_0 + z_1 + z_3)(y_0 + y_1 + y_3);
\end{aligned}$$

and

$$\Phi_{I3} \equiv Z_3(u)Y_3(u) \bmod P_3(u)$$
$$= (m_4 + m_5 + m_6 + m_8)u^2 + (m_4 + m_5 + m_7)u + m_4,$$

where

$$\begin{aligned}
m_4 &= z_0 y_0, \\
m_5 &= z_1 y_1, \\
m_6 &= z_2 y_2, \\
m_7 &= (z_0 + z_1)(y_0 + y_1), \\
m_8 &= (z_0 + z_2)(y_0 + y_2).
\end{aligned}$$

To reconstruct the length 7 aperiodic convolution, $\Phi(u)$, from the above length 6 convolution, we will require one additional multiplication, $m_9 = z_3 y_3$. The total multiplicative complexity can be further reduced, however, by noting that if we define the multiplications,

$$m_1' = (z_2 + z_3)(y_2 + y_3)$$

$$m_2' = (z_1 + z_3)(y_1 + y_3),$$

it becomes possible to write the coefficients of $\Phi_{I2}(u)$ as

$$(m_1 + m_3) = m_1' + m_2' + m_5 + m_6 + m_7 + m_8$$

$$(m_1 + m_2) = m_0 + m_2' + m_4 + m_6 + m_7 + m_9$$

and save one multiplication.

Reconstruction of the length 6 convolution may be carried out using (3.12 - 3.15), i.e.,

$$\Phi_I(u) \equiv \left[\sum_{i=1}^{3} S_i(u)\, \Phi_{Ii}(u) \right] \bmod P_I(u).$$

The reconstruction polynomials can be obtained using Euclid's algorithm; they are

$$S_1(u) = u^5 + u^4 + u^3, \quad S_2(u) = u^5 + u^4, \quad \text{and } S_3(u) = u^3 + 1,$$

giving the result

$$
\begin{aligned}
\Phi_I(u) \;=\; & (m_1' + m_6 + m_9)u^5 + (m_2' + m_5 + m_6 + m_9)u^4 \\
& + (m_0 + m_1' + m_2' + m_4 + m_5 + m_6 + m_7 + m_8)u^3 \\
& + (m_4 + m_5 + m_6 + m_8)u^2 + (m_4 + m_5 + m_7)u + m_4.
\end{aligned}
$$

Let the coefficients of $\Phi_I(u)$ be ϕ_{It}, $t = 0, 1, \ldots, 5$; the coefficients of $\Phi(u)$ are

$$
\phi_t(u) = \begin{cases}
\phi_{It}, & t = 0, 1, 2, 4, 5; \\
\phi_{It} + m_9, & t = 3; \\
m_9, & t = 6.
\end{cases}
$$

Directly from these expressions, we find that the NC algorithm for computing the system of bilinear forms Φ is given by $\phi = A^T(C^T \mathbf{z} \bullet B\mathbf{y})$, i.e.,

$$
\phi = \begin{bmatrix}
0 & 0 & 0 & 1 & 0 & 0 & 0 & 0 & 0 \\
0 & 0 & 0 & 1 & 1 & 0 & 1 & 0 & 0 \\
0 & 0 & 0 & 1 & 1 & 1 & 0 & 1 & 0 \\
1 & 1 & 1 & 1 & 1 & 1 & 1 & 1 & 1 \\
0 & 0 & 1 & 0 & 1 & 1 & 0 & 0 & 1 \\
0 & 1 & 0 & 0 & 0 & 1 & 0 & 0 & 1 \\
0 & 0 & 0 & 0 & 0 & 0 & 0 & 0 & 1
\end{bmatrix}
\left[\begin{bmatrix}
1 & 1 & 1 & 1 \\
0 & 0 & 1 & 1 \\
0 & 1 & 0 & 1 \\
1 & 0 & 0 & 0 \\
0 & 1 & 0 & 0 \\
0 & 0 & 1 & 0 \\
1 & 1 & 0 & 0 \\
1 & 0 & 1 & 0 \\
0 & 0 & 0 & 1
\end{bmatrix}
\begin{bmatrix} z_0 \\ z_1 \\ z_2 \\ z_3 \end{bmatrix} \right]
$$

$$
\bullet \left[\begin{bmatrix}
1 & 1 & 1 & 1 \\
0 & 0 & 1 & 1 \\
0 & 1 & 0 & 1 \\
1 & 0 & 0 & 0 \\
0 & 1 & 0 & 0 \\
0 & 0 & 1 & 0 \\
1 & 1 & 0 & 0 \\
1 & 0 & 1 & 0 \\
0 & 0 & 0 & 1
\end{bmatrix}
\begin{bmatrix} y_0 \\ y_1 \\ y_2 \\ y_3 \end{bmatrix} \right]
$$

and has a total of 9 multiplications over the field of constants $GF(2)$.

The above length 7 aperiodic convolution design over the field of constants $GF(2)$ has led to a bilinear algorithm requiring 9 multiplications. We believe this is the *lowest* number possible over $GF(2)$; attempts to design more efficient algorithms will result in certain elements of the matrices A, B, and C *not* being in $GF(2)$. Over a field (still finite), but large enough, i.e., $|\mathcal{F}| \geq k+d-2$, the minimum number of multiplications is $\mu = k + d - 1 = 7$; this follows from the modified Toom-Cook algorithm of Section 3.2.1. There are two points which should be made: first, an algorithm of this type achieves its efficiency not only by applying the notion of coefficient wraparound, but also by identifying certain multiplications, the number of which can be reduced (at the expense, usually, of an increase in additions) and, second, once good algorithms of this type are constructed, they can be *iterated* (using disjoint sets of indeterminates) to obtain good algorithms of larger size. For the above example, iterating the algorithm once results in an algorithm for computing the aperiodic convolution of two sequences of length $2^r = 8$ having a multiplicative complexity of $3^r = 27$ multiplications. This is an excellent way of obtaining efficient aperiodic convolution algorithms for digital signal processing applications.

As discussed in Chapter 2, a computation of the above type can be associated with the construction of a linear code. In fact, the matrix C in the above NC algorithm can be thought of as the generator matrix of a $(9, 4, 4)$ linear code where the multiplicative complexity determines the code length. If we wish to obtain close-to-optimal linear codes of higher dimension, it may be better to make a fresh start rather than to use the process of iteration, albeit the former is a more difficult undertaking. Say we wish to design a linear code of dimension $k = 8$ and minimum distance $d = 8$. We choose $s = 2$ and the factors $P_1(u) = u+1$, $P_2(u) = u^2+u+1$, $P_3(u) = u^3+u^2+u+1$, and $P_4(u) = u^4$ (note that the multiplication of degree 3 polynomials mod u^4 only requires 8 multiplications); this results in a multiplicative complexity of 26 and the $(26, 8, 8)$ linear code over $GF(2)$. It is quite crucial to realize that the multiplicative complexity of an $N = k + d - 1$ point aperiodic convolution determines the code length; in this sense, a given value of N is associated with a *linear code family* for which $k + d$ is constant. As an example, the $(26, 8, 8)$, $(26, 9, 7)$, $(26, 7, 9)$, and $(26, 5, 11)$ linear codes are all members of the same family, and their generator matrices are all simply interrelated.

Finally, we mention the approach to bilinear algorithm design suggested in [3.37]. There, the concern is chiefly with the multiplicative complexity of arbitrarily long length aperiodic convolution when $\mathcal{F}$ is a finite field. To examine this, the authors propose a recursive (iterative) CRT-based method in which polynomial factors are selected from a list of monic *irreducible* polynomials with coefficients from $GF(p)$. Over $GF(2)$ and $GF(3)$, these polynomials appear in brackets in Appendix C. The approach is systematic, but not quite as flexible (the only assumption we require is that the polynomials be mutally prime) or as efficient ($k = d = 4$ requires 11 multiplications; consequently, $k = d = 8$ requires 33 multiplications) as the method outlined here. Of course, in an asymptotic study of the type found in [3.37], there is a need for exactly this type of systematic approach; the authors have developed very efficient bilinear algorithms elsewhere. An important point is that the bounds on complexity, say when $|\mathcal{F}| = p$, obtained in [3.37] apply as well to the design procedure developed here. The study of these complexity bounds and, perhaps, the derivation of even sharper results is an open and indeed fascinating area of research.

3.3 References

[**3.1**] R. C. Agarwal and J. W. Cooley, "New Algorithms for Digital Convolution," *IEEE Trans. Acoust., Speech, Signal Processing*, vol. ASSP-25, pp. 392-410, Oct. 1977.

[**3.2**] R. C. Agarwal and J. W. Cooley, "New Algorithms for Digital Convolution", in Proc. *IEEE Int. Conf. Acoust., Speech, Signal Processing*, Hartford, CT, pp. 360-362, May 1977.

[**3.3**] S. Winograd, "On Computing the Discrete Fourier Transform," *Math. of Comput.*, vol. 32, pp. 175-199, Jan. 1978.

[**3.4**] R. C. Agarwal and C. S. Burrus, "Fast Convolution Using Fermat Number Transforms with Applications to Digital Filtering," *IEEE Trans. Acoust., Speech, Signal Processing*, vol. ASSP-22, pp. 87-99, Apr. 1974.

[**3.5**] R. C. Agarwal and C. S. Burrus, "Number Theoretic Transforms to Implement Fast Convolutions," *Proc. IEEE*, vol. 63, pp. 550-560, Apr. 1975.

[**3.6**] C. M. Rader, "Discrete Convolutions via Mersenne Transform," *IEEE Trans. Comput.*, vol. C-21, pp. 1269- 1273, Dec. 1972.

[**3.7**] I. S. Reed and T. K. Truong, "The Use of Finite Fields to Compute Convolutions," *IEEE Trans. Inform. Theory*, vol. IT-21, pp. 208-213, Mar. 1975.

[**3.8**] I. S. Reed and T. K. Truong, "Complex Integer Convolutions over a Direct Sum of Galois Fields," *IEEE Trans. Inform. Theory*, vol. IT-21, pp. 657-661, Nov. 1975

[**3.9**] N. S. Reddy and V. U. Reddy, "On Convolutional Algorithms for Small Word Length Digital Filtering Applications," *IEE J. Electron., Circuits, Syst.*, vol. 3, pp. 253-256, Nov. 1979.

[**3.10**] N. S. Reddy and V. U. Reddy, "Combination of Complex Rectangular Transforms and FNT to Implement Fast Convolution," *Electron. Lett.*, vol. 15, pp. 716-717, Oct. 1979.

[**3.11**] V. U. Reddy and N. S. Reddy, "Complex Rectangular Transforms," in Proc. *IEEE Int. Conf. Acoust., Speech, Signal Processing*, Washington, DC, pp. 518-521, Apr. 1979.

[**3.12**] H. J. Nussbaumer and P. Quandalle, "Computations of Convolutions and Discrete Fourier Transforms by Polynomial Transforms," *IBM J. Res. Develop.*, vol. 22, pp. 134- 144, Mar. 1978.

[**3.13**] H. J. Nussbaumer, "Complex Convolutions via Fermat Number Transforms," *IBM J. Res. Develop.*, vol. 22, pp. 282-284, Dec. 1976.

[**3.14**] H. J. Nussbaumer, "Fast Polynomial Transform Algorithms for Digital Convolution," *IEEE Trans. Acoust., Speech, Signal Processing*, vol. ASSP-28, pp. 205-215, Apr. 1980.

[**3.15**] J. H. McClellan and C. M. Rader, *Number Theory in Digital Signal Processing*, Prentice Hall, 1979.

[**3.16**] H. J. Nussbaumer, *Fast Fourier Transforms and Convolution Algorithms*, Springer-Verglag, 1981.

[**3.17**] M. D. Wagh and S. D. Morgera, "A New Structured Design Method for Convolutions over Finite Fields, Part I," *IEEE Trans. Inform. Theory*, vol. IT-29, pp. 583-595, July 1983.

[**3.18**] S. D. Morgera and M. D. Wagh, "Bilinear Cyclic Convolution Algorithms over Finite Fields," *IEEE Int. Symp. Inform. Theory*, Santa Monica, CA, Feb. 1981.

[**3.19**] P. C. Balla, A. Antoniou, and S. D. Morgera, "Higher-Radix Aperiodic-Convolution Algorithms,' *IEEE Trans.Acoust.,Speech, Signal Processing*, vol. ASSP-34, pp. 60-68, Feb. 1986.

[**3.20**] H. Krishna, "Computational Complexity of Bilinear Forms - Algebraic Coding Theory and Applications to Digital Communications Systems," Thesis in the Dept. of Elect. Engin., Concordia University, Montreal, Canada, Aug. 1985.

[**3.21**] E. R. Berlekamp, *Algebraic Coding Theory*, McGraw-Hill, 1968.

[**3.22**] S. W. Golomb, *Shift Register Sequences*, Holden-Day, San Francisco, CA, 1967.

[**3.23**] J. M. Pollard, "The Fast Fourier Transform in a Finite Field," *Math. of Comput.*, vol. 25, pp. 365-374, Apr. 1971.

[**3.24**] S. D. Morgera, "Efficient Synthesis and Implementation of Large Discrete Fourier Transformations," *SIAM J. Comput.*, vol. 9, pp. 251-272, May 1980.

[**3.25**] S. D. Morgera, "Prime Power Synthesis of Linear Transformations," *IEEE Trans. Acoust., Speech, Signal Processing*, vol. ASSP-28, pp. 252-254, Apr. 1980.

[**3.26**] C. M. Rader, "Discrete Fourier Transforms when the Number of Data Samples is Prime", *Proc. IEEE*, vol. 56, pp. 1107-1108, June 1968.

[**3.27**] S. Winograd, "On the Multiplicative Complexity of the Discrete Fourier Transform," *Adv. in Math.*, vol. 32, pp. 83-117, 1979.

[**3.28**] F. P. Preparata and D. V. Sarwate, "Computational Complexity of Fourier Transforms over Finite Fields," Univ. of Illinois *Co-ord. Sci. Lab., Rept. No. R-745*, UILU-ENG-76-2233, 1976.

[**3.29**] S. Winograd, "Some Bilinear Forms Whose Multiplicative Complexity Depends on the Field of Constants," *Math. Syst. Theory*, vol. 10, pp. 169-180, 1977.

[**3.30**] I. S. Reed, T. K. Truong, R. L. Miller, and B. Benjauthrit, "Further Results on Fast Transforms for Decoding Reed-Solomon Codes over $GF(2^m)$ for $m = 4, 5, 6, 8$," in *Deep Space Network Prog. Rept. 42-50*, Jet. Prop. Lab., Pasadena, CA, pp. 132-154, Jan. 1979.

[**3.31**] C. M. Fiduccia and Y. Zalcstein, "Algebras Having Minimum Multiplicative Complexities," *Tech. Rept. 46*, Dept. of Comp. Sci., State Univ. of NY at Stony Brook, Aug. 1975.

[**3.32**] A. L. Toom, "The Complexity of a Scheme of Functional Elements Realizing the Multiplication of Integers," Soviet Math. Dokl., vol. 4, pp. 714-716, 1963.

[**3.33**] D. E. Knuth, *The Art of Computer Programming, vol. 2, Seminumerical algorithms*, Addison-Wesley, 1969.

[**3.34**] R. C. Agarwal and C. S. Burrus, "Fast One Dimensional Convolution by Multidimensional Techniques," *IEEE Trans. Acoust., Speech, Signal Processing*, vol. ASSP-22, pp. 1-10, Feb. 1974.

[**3.35**] W. W. Peterson and E. J. Weldon, Jr., *Error-Correcting Codes*, The M.I.T. Press, Cambridge, MA, 1972.

[**3.36**] J. D. Alanen and D. E. Knuth, "Tables of Finite Fields," *Sankhya (Calcutta)*, vol. 26, pp. 305-328, 1964.

[**3.37**] A. Lempel, G. Seroussi, and S. Winograd, "On the Complexity of Multiplication in Finite Fields," *Theo. Comp. Sci.*, vol. 22, pp. 285-296, 1983.

3.4 Appendix A

Proof of Properties $P1 - P5$

The reader may consult [3.21] for additional insight regarding the construction used; in particular, see Section 4.4 of that reference concerning the algebraic structure of finite fields. By way of introduction, let $N = p^n - 1$ and consider the correspondence of the integer set $\{0, 1, \cdots, N - 1\}$ with the elements of $GF(p^n)$ as $i \leftrightarrow \alpha^i$, where α is a primitive element of $GF(p^n)$. Due to the manner in which the cyclotomic sets S_i are defined, $j_1, j_2 \epsilon S_i$, $j_1 \neq 0$; $j_2 \neq 0$, yield the conjugates α^{j_1} and α^{j_2}. Thus, σ_i equals the number of conjugates of α^i and $i \epsilon S$ iff it is the smallest exponent of α amongst all the conjugates of α^i. The proof of $P1$ is obvious; we start with $P2$.

PROOF (P2). For any integer $t \mid n$, $GF(p^t) \subseteq GF(p^n)$, and the element α^θ generates the multiplicative group of $GF(p^t)$ when $\theta = (p^n - 1)/(p^t - 1)$. Every nonzero element of $GF(p^t)$ is of the form $\alpha^{\theta \ell}$ for ℓ a positive integer. Since all the conjugates of α^θ belong to $GF(p^t)$, we have $j > \theta$ for any conjugate α^j; thus, $\theta \epsilon S$. Moreover, α^θ generates the multiplicative group of $GF(p^t)$, implying that it has t conjugates and, therefore, $\sigma_\theta = t$.

PROOF (P3). For any $i \epsilon S$, $i \neq 0$, α^i must belong to some smallest field, $GF(p^t) \subseteq GF(p^n)$, in the sense that α^i does not belong to any subfield of $GF(p^t)$. With $\theta = (p^n - 1)/(p^t - 1)$, $\alpha^i = \alpha^{\theta \ell}$, or $i = \theta \ell$, for some positive integer ℓ. Moreover, since α^i does not belong to any subfield of $GF(p^t)$, it has t conjugates; thus, $\sigma_i = t = \sigma_\theta$.

The properties $P4$ and $P5$ follow directly from the above.

3.5   Appendix B

Bilinear Polynomial Multiplication Algorithms

The interested reader may also consult [3.1, 3.34] for some of the same (after adaptation to a field of constants $GF(p)$), or slightly different, bilinear algorithms. Those provided here are valid over any field. In this appendix, we use the indeterminate input sequences $\{x_0, x_1, \cdots\}$ and $\{y_0, y_1, \cdots\}$ and the output sequence $\{\psi_0, \psi_1, \cdots\}$. Multiplications are designated $m_0, m_1, \cdots$.

DEGREE 0.

$$\psi_0 = x_0 y_0.$$

Complexity: 1 multiplication.

DEGREE 1.

$$\psi(u) = \left(\sum_{i=0}^{1} x_i u^i \right) \left(\sum_{i=0}^{1} y_i u^i \right) = \sum_{i=0}^{2} \psi_i u^i.$$

Let

$$\begin{aligned} m_0 &= x_0 y_0, \\ m_1 &= x_1 y_1, \\ m_2 &= (x_0 + x_1)(y_0 + y_1); \end{aligned}$$

then

$$\begin{aligned} \psi_0 &= m_0, \\ \psi_1 &= -m_0 - m_1 + m_2, \\ \psi_2 &= m_1. \end{aligned}$$

Complexity: 3 multiplications.

DEGREE 2.

$$\psi(u) = \left(\sum_{i=0}^{2} x_i u^i \right) \left(\sum_{i=0}^{2} y_i u^i \right) = \sum_{i=0}^{4} \psi_i u^i.$$

Let

$$
\begin{aligned}
m_0 &= x_0 y_0, \\
m_1 &= x_1 y_1, \\
m_2 &= x_2 y_2, \\
m_3 &= (x_0 + x_1)(y_0 + y_1), \\
m_4 &= (x_1 + x_2)(y_1 + y_2), \\
m_5 &= (x_0 + x_2)(y_0 + y_2);
\end{aligned}
$$

then

$$
\begin{aligned}
\psi_0 &= m_0, \\
\psi_1 &= -m_0 - m_1 + m_3, \\
\psi_2 &= -m_0 + m_1 - m_2 + m_5, \\
\psi_3 &= -m_1 - m_2 + m_4, \\
\psi_4 &= m_2.
\end{aligned}
$$

Complexity: 6 multiplications.

(Note: $\psi(u) \bmod u^3$ requires only 5 multiplications; delete m_5, ψ_3, ψ_4 in the above and replace m_5 by m_4 in ψ_2).

DEGREE 3.

$$
\psi(u) = \left(\sum_{i=0}^{3} x_i u^i \right) \left(\sum_{i=0}^{3} y_i u^i \right) = \sum_{i=0}^{6} \psi_i u^i.
$$

Let

$$
\begin{aligned}
m_0 &= x_0 y_0, \\
m_1 &= x_1 y_1, \\
m_2 &= (x_0 + x_1)(y_0 + y_1), \\
m_3 &= x_2 y_2, \\
m_4 &= x_3 y_3, \\
m_5 &= (x_2 + x_3)(y_2 + y_3), \\
m_6 &= (x_0 + x_2)(y_0 + y_2), \\
m_7 &= (x_1 + x_3)(y_1 + y_3), \\
m_8 &= (x_0 + x_1 + x_2 + x_3)(y_0 + y_1 + y_2 + y_3);
\end{aligned}
$$

then

$$\psi_0 = m_0,$$
$$\psi_1 = -m_0 - m_1 + m_2,$$
$$\psi_2 = -m_0 + m_1 - m_3 + m_6,$$
$$\psi_3 = m_0 + m_1 - m_2 + m_3 + m_4,$$
$$-m_5 - m_6 - m_7 + m_8,$$
$$\psi_4 = -m_1 + m_3 - m_4 + m_7,$$
$$\psi_5 = -m_3 - m_4 + m_5,$$
$$\psi_6 = m_4.$$

Complexity: 9 multiplications.

(Note: $\psi(u)\bmod u^4$ requires only 8 multiplications; delete m_8 and ψ_i, $i = 4, 5, 6$, in the above and replace m_2 by m_4 in ψ_1, $m_1 - m_3 + m_6$ by $-m_2 + m_5$ in ψ_2, and delete $m_4 - m_5$ in ψ_3).

3.6 Appendix C

Some Monic Polynomials Over $GF(2)$ and $GF(3)$

The list is constructed so that all polynomials of degree ℓ are not products of distinct polynomials of degrees $< \ell$. The monic irreducible polynomials over $GF(2)$ and $GF(3)$ are indicated using brackets. Of course, the reciprocal polynomial of an irreducible polynomial is also irreducible; both the irreducible polynomial and its reciprocal appear in brackets.

Over $GF(2)$

DEGREE 1: $[u]$, $[u + 1]$
DEGREE 2: u^2, $u^2 + 1$, $[u^2 + u + 1]$
DEGREE 3: u^3, $u^3 + u^2 + u + 1$, $[u^3 + u + 1]$, $[u^3 + u^2 + 1]$
DEGREE 4: u^4, $u^4 + 1$, $u^4 + u^2 + 1$, $[u^4 + u + 1]$,
                $[u^4 + u^3 + u^2 + u + 1]$, $[u^4 + u^3 + 1]$
DEGREE 5: u^5, $u^5 + u^4 + u + 1$, $[u^5 + u^2 + 1]$,
                $[u^5 + u^4 + u^3 + u^2 + 1]$, $[u^5 + u^4 + u^2 + u + 1]$,
                $[u^5 + u^3 + u^2 + u + 1]$, $[u^5 + u^4 + u^3 + u + 1]$,
                $[u^5 + u^3 + 1]$.

Over $GF(3)$

DEGREE 1: $[u]$, $[u + 1]$, $[u + 2]$
DEGREE 2: u^2, $[u^2 + 1]$, $u^2 + u + 1$, $[u^2 + u + 2]$, $[u^2 + 2u + 2]$,
                $u^2 + 2u + 1$
DEGREE 3: u^3, $u^3 + 1$, $u^3 + 2$, $[u^3 + 2u + 1]$, $[u^3 + 2u + 2]$,
                $[u^3 + u^2 + 2]$, $[u^3 + u^2 + u + 2]$, $[u^3 + u^2 + 2u + 1]$,
                $[u^3 + 2u^2 + 1]$, $[u^3 + 2u^2 + u + 1]$,
                $[u^3 + 2u^2 + 2u + 2]$.

Chapter 4

Multidimensional Methods

In this chapter, we are interested in the design and study of long length cyclic convolutions and some preliminary results on their relation to linear codes. We have already come across one approach for long lengths (although, in the context of causal aperiodic convolution); this is the method discussed in Section 3.2.2 and described in [4.1]. One possible disadvantage of this approach is that the mapping of the one-dimensional sequence into multidimensional sequences is *lexicographic* and, therefore, the resulting multidimensional array is cyclic in only one dimension. This occurs, of course, due to the fact that the only requirement imposed is that the overall length be composite with integer factors.

The work in [4.2] extended and improved the multidimensional mapping by requiring that the overall length be composite with *mutually prime* factors, thereby resulting in a multidimensional array cyclic in all dimensions. In this approach, the Chinese Remainder Theorem is used to map the sequences into the multidimensional arrays (see [4.3 - 4.4] for an application of this to the DFT). An obvious advantage is that longer lengths can systematically be accomodated in mutually prime multiples. This method was cast in a formal light by complexity results for systems of bilinear forms defined by the product of two polynomials modulo a third polynomial [4.5]. The idea of designing long length convolutions using mutually prime factors and the Chinese Remainder Theorem provides, in some sense, a basis for comparison in this chapter. When the cyclic convolution length, $N = \prod_{i=1}^{r} N_i$, $(N_i, N_j) = 1$, $i, j = 1, 2, \ldots, r$; $i \neq j$, the CRT-based multidimensional methods have a multiplicative complexity equal to $\tilde{M}(N) = \prod_{i=1}^{r} \tilde{M}(N_i)$, where $\tilde{M}(\cdot)$ denotes the number of multiplications required for a cyclic convolution of the indicated length. The number of additions depends on the order in which the subproblems are computed. We shall call algorithms in this CRT-based class *mutually prime factor algorithms*.

Of course, in the same manner, it is possible to construct long cyclic convolutions from shorter, mutually prime length cyclic convolution algorithms designed as in Section 3.1, but is this better than designing the long length algorithm *directly*? We answer this question in this section by showing that the multiplicative complexity $M(N_1 N_2)$ for long length cyclic convolution algorithms over $GF(p)$ designed according to Section 3.1 is *less than or equal to* $M(N_1)M(N_2)$, the product of the multiplicative complexities of the individual algorithms designed in the same manner. For assumptions, we only require the usual $(N, p) = 1$ design requirement, where $N = N_1 N_2$, in addition to $(N_1, N_2) = 1$. Furthermore, we prove the strict inequality $M(N_1 N_2) < M(N_1)M(N_2)$, unless the orders of p mod N_i, $i = 1, 2$, are also mutually prime, in which case, equality holds. The long length direct-design algorithms also turn out to be more efficient than the mutally prime factor algorithms of [4.2].

We consider the results of this chapter to be of some significance, as they have important implications, both for efficient long length cyclic convolution and DFT algorithm design (note that due to the systematic approach to design taken in Section 3.1, it is entirely practical to directly design cyclic convolution algorithms for long lengths) and for understanding the complexity of cylic convolution over a finite field when the length N is arbitrarily large.

We know that over a sufficiently large field (such as assumed in the Toom-Cook procedure), the minimum number of multiplications is $\mu = 2N - 1$ for degree $(N - 1)$ polynomial multiplication, followed by reduction modulo a monic irreducible polynomial of degree N. When the field $GF(p)$ is small, the length N is large, and we permit the modulo reduction to be performed with respect to a polynomial of degree N having a number of irreducible factors over $GF(p)$, it is possible to probabilistically argue that the number of multiplications has, on the average, the lower bound, $2N - \log N$. Of importance also is the upper bound on the number of multiplications under the same conditions and whether it is linear in N, i.e., $M(N) \leq cN$, where c is a positive constant and is independent of N. The direct method of this section and of [4.6] appears to show promise in taking a first step towards achieving such a linear upper bound. The existence of such a bound would be of considerable interest to coding theorists as well as those seeking efficient digital signal processing algorithms.

4.1 Long Length Cyclic Convolution

We address the question of the computation of the length N cyclic convolution $\psi = \mathbf{x} * \mathbf{y}$, where $\mathbf{x}$ and $\mathbf{y}$ are vectors over $GF(p^m)$ and the field of constants is $GF(p)$. In Section 3.1, we developed a bilinear algorithm for the computation of ψ; the multiplicative complexity (3.5) of the algorithm is

$$M(N) = 1 + \sum_{i \in S_N} M_{\sigma_i}. \tag{4.1}$$

The quantity M_{σ_i} denotes the multiplicative complexity of degree $(\sigma_i - 1)$ polynomial multiplication, and S_N is the index set associated with cyclotomic sets resulting from a partitioning of an integer set. At first, the cyclotomic set construction assumed $N = p^n - 1$ (for which S was used to denote the index set) and was then extended in Section 3.1.4 to all N such that $(N, p) = 1$, provided $N | (p^n - 1)$. In the latter case, a more direct approach is taken below to the cyclotomic set construction or, equivalently, the isolation of the roots of the irreducible factors of $u^N - 1$ over $GF(p)$.

DEFINITION 1. Assume $(N, p) = 1$. The *order* of p mod N is the smallest integer e such that $N | (p^e - 1)$, or, equivalently, such that $p^e \equiv 1 \bmod N$. We shall interchangeably refer to e as the *exponent* of N with respect to p. As an example, with respect to $p = 2$, the exponents of $N = 3, 5, 9, 35$ are $e = 2, 4, 6, 12$, respectively.

We now define the cyclotomic set construction to be used here. Rather than choose new notation, we keep the notation the same as the earlier cyclotomic set notation.

DEFINITION 2. Assume $(N, p) = 1$ and let e denote the order of p mod N. Partition the set of integers $\{j(p^e - 1) / N : 1 \le j \le N - 1\}$

into sets S_{i_1}, S_{i_2},.... A *cyclotomic set* S_i is generated by starting from the smallest integer i not covered in an earlier set and computing other members as

$$S_i = \left\{ i,\ ip \bmod (p^e - 1),\ ip^2 \bmod (p^e - 1), \ldots \right\}.$$

We denote the cardinality of S_i, $|S_i|$, by σ_i and the set of indices $\{i_1, i_2, \ldots\}$ by S_N. Some properties associated with this construction are

P1') $0 \notin S_N$ (by construction);

P2') $\sigma_i | e$ for all $i \in S_N$;

P3')   For any N, $\sigma_i = e$, at least for one $i \in S_N$;

P4')   For N prime, $\sigma_i = e$, for all $i \in S_N$;

P5') $\sum_{i \in S_N} \sigma_i = N - 1$;

P6')   If $N = q^n$, where $q \neq p$ is a prime, then any σ_i, $i \in S_N$, has the form $\sigma_i = e' q^\ell$, where e' is the order of p mod q and ℓ depends on i, $0 \leq \ell \leq n - 1$.

Proofs of the above properties may be found in Appendix A. The following example illustrates the above construction.

EXAMPLE 1.   Consider the field of constants $GF(2)$ and $N = 3^2$. The exponent e of $N = 9$ with respect to $p = 2$ is the smallest integer such that $N | (p^e - 1)$; we find that $e = 6$. We partition the integer set $\{ j(p^e - 1) / N : 1 \leq j \leq N - 1 \} = \{7, 14, 21, 28, 35, 42, 49, 56\}$ into the sets

$$\begin{aligned} S_7 &= \{7, 14, 28, 56, 49, 35\} \text{ and} \\ S_{21} &= \{21, 42\}. \end{aligned}$$

We have $\sigma_7 = 6$, $\sigma_{21} = 2$, and $S_9 = \{7, 21\}$. Note that σ_7 and σ_{21} divide e and that $\sigma_7 = e$, in accord with P2' and P3', respectively. Property P6' is also satisfied, as $e' = 2$ and $\sigma_7 = e' q^\ell$, for $\ell = 1$; $\sigma_{21} = e' q^\ell$, for $\ell = 0$. As another example, the reader should refer to that of Section

3.1.4, where $N = 4$ and $S_4 = \{2, 4\}$ over $GF(3)$.

There is one additional property of the above construction that is quite crucial to obtaining multiplicative complexity results for long lengths. To present this property, we first assume that $N = N_1 N_2$, where $(N_1 N_2) = 1$. Let $\sum = \{\sigma_i : i \epsilon S_N\}$ and $\sum_j = \{\sigma_{i_j} : i_j \epsilon S_{N_j}\}$, $j = 1, 2$. Property *P7'* permits a *unique* reconstruction of $\sum$ from the $\sum_i$, $i = 1, 2$:

P7')   If $N = N_1 N_2$, $(N_1, N_2) = 1$, $\sum = \sum' \cup \sum_1 \cup \sum_2$, where, for every pair of elements, $\sigma_{i_1} \epsilon \sum_1$, $\sigma_{i_2} \epsilon \sum_2$, $\sum$ has $\gcd(\sigma_{i_1}, \sigma_{i_2})$ occurrences of $\mathrm{lcm}(\sigma_{i_1}, \sigma_{i_2})$ (note: we use (a, b) and $\gcd(a, b)$ interchangeably to denote the greatest common divisor of two integers a and b, with the abbreviated form preferred, if it is certain that no confusion will arise). The notation $\mathrm{lcm}(\cdot, \cdot)$ denotes the least common multiple of the indicated quantities.

EXAMPLE 2.   Consider the field of constants $GF(2)$ and $N = 15$, $N_1 = 3$, and $N_2 = 5$. It is easy to verify that $S_3 = \{1\}$ and $S_5 = \{3\}$ with $\sum_1 = \{2\}$ and $\sum_2 = \{4\}$. In this simple case, $\sum'$ has $\gcd(2, 4)$ occurrences of $\mathrm{lcm}(2, 4)$; thus,

$$\sum' = \{4, 4\} \text{ and}$$
$$\sum = \{2, 4, 4, 4\}.$$

The elements of $\sum$ so obtained represent the correct cardinalities for the sets $S_1 = \{1, 2, 4, 8\}$, $S_3 = \{3, 6, 12, 9\}$, $S_5 = \{5, 10\}$, and $S_7 = \{7, 14, 13, 11\}$ associated with the length $N = 15$. The reader may recognize *P7'* as a result from the *theory of cyclotomy* for sorting out the multiplicative orders of $p \bmod N_i$.

Before proceeding to results of a general nature, we show how the above properties may be applied in a specific case. Assume N prime; using *P4'* and *P5'*, we have $\sigma_i = e$, for all $i \epsilon S_N$, and $|S_N| = (N - 1)/e$, where e is the order of $p \bmod N$. Using (4.1), this immediately results in

$$M(N) = 1 + (N - 1) M_e/e \quad (N \text{ prime}). \tag{4.2}$$

As an example, take the cyclic convolution of length $N = 17$ sequences over the field of constants $GF(2)$. For $N = 17$, we have $e = 8$ and $M_e = 27$ multiplications (iterate the algorithm of Section 3.2.3 designed for degree 3 polynomial multiplication); thus $M(17) = 1 + 2M_e = 55$ multiplications. The theoretical minimum, $\mu \geq 2N - |S_N| - 1 = 31$ multiplications.

We now consider the construction of a cyclic convolution of length $N = N_1 N_2$, where $(N_1, N_2) = 1$, in two ways:

1. Designing cyclic convolutions of length N_1 and N_2 *separately* using the method of Section 3.1 and then constructing the cyclic convolution of length N and

2. Designing a cyclic convolution of length N *directly* using the method of Section 3.1.

In the case of mutually prime factor algorithms, it is not difficult to show that the algorithm for length N sequences having two mutually prime factors N_1 and N_2 takes the form

$$\psi = (C^{(1)} \otimes C^{(2)}) \left[(A^{(2)} \otimes A^{(1)})\mathbf{x} \bullet (B^{(2)} \otimes B^{(1)})\mathbf{y} \right], \tag{4.3}$$

where $\otimes$ denotes the Kronecker (direct) product and the bilinear algorithms, $\psi_i = C^{(i)}(A^{(i)}\mathbf{x}_i \bullet B^{(i)}\mathbf{y}_i)$, are associated with the factors N_i, $i = 1, 2$, respectively. The matrices $A^{(i)}$, $B^{(i)}$, and $C^{(i)}$ are $(\tilde{M}(N_i) \times N_i)$, $(\tilde{M}(N_i) \times N_i)$, and $(N_i \times \tilde{M}(N_i))$-dimensional, respectively; consequently, the multiplicative complexity of the approach is $\tilde{M}(N_1)\tilde{M}(N_2)$. The form (4.3) follows directly from the fact that $\Psi(u) = X(u)Y(u)$ mod $(u^N - 1)$ is equivalent to $\Psi(u, v) = X(u, v)Y(u, v)$ mod $(u^{N_1} - 1)$ mod $(v^{N_2} - 1)$. This equivalence is obtained simply by renaming the variables (generating polynomial coefficients) in accord with the Chinese Remainder Theorem. This is the approach taken (over the rationals) in [4.2] and other references for multifactor design. As an example, when $N = 15$, $N_1 = 3$, and $N_2 = 5$, the above construction leads to 40 multiplications over the field of constants $GF(2)$; whereas, direct design results in $M(15) = 31$ multiplications (cf. Table 3.1). The implication, on a deeper level, is that certain cyclotomic sets *split* in the above example in forming the Nth roots of unity from products of the N_1th and N_2th roots of unity. More will be said regarding this in the sequel.

4.1.1 Mutual Prime Factor versus Direct Designs

The central result of this section concerns a comparison of the multiplicative complexity $M(N_1N_2)$ versus $M(N_1)M(N_2)$, for $(N, p) = 1$; $(N_1, N_2) = 1$. The result will be seen to depend on a relation between the orders of $p \bmod N_i, i = 1, 2$. To set the notation, let $p^{e_1} \equiv 1 \bmod N_1$ and $p^{e_2} \equiv 1 \bmod N_2$. In the following lemma and proof, it is important to recall that $(N_1, N_2) = 1$ does *not* imply $\gcd(e_1, e_2) = 1$ and, in fact, we may have $\gcd(e_1, e_2) > 1$. For example, for $N_1 = 7$, $N_2 = 5$, $\gcd(e_1, e_2) = 1$; whereas, for $N_1 = 13$, $N_2 = 9$, $\gcd(e_1, e_2) = 6$. We now state an intermediate result.

LEMMA 4.1 *Let the length of the cyclic convolution over $GF(p)$ be $N = N_1N_2$, where $(N_1, N_2) = 1$, $(N, p) = 1$, and e_i are the orders of $p \bmod N_i$, $i = 1, 2$; then, for the cyclic convolution algorithm having multiplicative complexity given by (4.1),*

$$
\begin{aligned}
M(N_1N_2) &= M(N_1) + M(N_2) - 1 \\
&+ \sum_{i_1 \in S_{N_1}} \sum_{i_2 \in S_{N_2}} [\gcd(\sigma_{i_1}, \sigma_{i_2})] \, M_{lcm\,(\sigma_{i_1}, \sigma_{i_2})}.
\end{aligned} \quad (4.4)
$$

PROOF. From (4.1) we have

$$
M(N_1N_2) = 1 + \sum_{i \in S_{N_1 N_2}} M_{\sigma_i}.
$$

Since $(N_1, N_2) = 1$, we may use $P7'$ to decompose the index set $S_{N_1 N_2}$ into S_{N_1}, such that for $i_1 \in S_{N_1}$, $\sigma_{i_1} \in \Sigma_1$; S_{N_2} such that for $i_2 \in S_{N_2}$, $\sigma_{i_2} \in \Sigma_2$; and a remaining *interaction term* creating an appropriate sequence so as to cover all $i \in S_{N_1 N_2}$, $\sigma_i \in \Sigma$. This decomposition provides us with

$$
\begin{aligned}
M(N_1N_2) &= 1 + \sum_{i_1 \in S_{N_1}} M_{\sigma_{i_1}} + \sum_{i_2 \in S_{N_2}} M_{\sigma_{i_2}} \\
&+ \sum_{i_1 \in S_{N_1}} \sum_{i_2 \in S_{N_2}} [\gcd(\sigma_{i_1}, \sigma_{i_2})] \, M_{lcm\,(\sigma_{i_1}, \sigma_{i_2})},
\end{aligned}
$$

where $M_{\text{lcm}\,(\sigma_{i_1},\sigma_{i_2})}$ is the multiplicative complexity for degree [lcm $(\sigma_{i_1},\sigma_{i_2}) - 1$] polynomial multiplication. Once again using (4.1) for the first two summations, we obtain the desired result.

In order to simplify, and even understand, the expression for the multiplicative complexity $M(N_1 N_2)$ obtained in Lemma 4.1, we must consider some of the various approaches discussed in Chapter 3 for aperiodic convolution, or polynomial multiplication, the multiplicative complexity of which is so important to the overall efficiency. Several basic approaches were discussed in Section 3.2; one involved extending the sequence length by padding zeroes and performing the cyclic convolution of the extended sequences. For a polynomial product where the indeterminate sequence length is $\text{lcm}(\sigma_{i_1},\sigma_{i_2})$, the complexity would be $M(2\text{lcm}(\sigma_{i_1},\sigma_{i_2}) - 1)$ using the cylic convolution algorithm of Section 3.1. A problem with this approach is that $\gcd(p, \text{lcm}(\sigma_{i_1},\sigma_{i_2})) = 1$ is required for algorithm design and difficult to ensure in practice (see previous example). A more promising approach is to *iterate* (recursively apply) small, efficient polynomial multiplication algorithms in order to achieve an algorithm for higher-degree polynomial multiplication. This method was described in Section 3.2.2 and only requires that there be a common integer factor in the lengths of the two indeterminate sequences to be convolved. A complete and efficient procedure for aperiodic convolution would also include the CRT and wraparound method of Section 3.2.3, wherein each subproblem of the form $\Phi_i(u) \equiv Z_i(u)Y_i(u) \bmod P_i(u)$ is computed using the above iterated algorithms.

To demonstrate the applicability of iterated algorithms to the present problem, we return to the problem of computing the polynomial product of length $\text{lcm}(\sigma_{i_1},\sigma_{i_2})$ sequences, $i_1 \epsilon S_{N_1}$, $i_2 \epsilon S_{N_2}$. By definition, we have

$$\text{lcm}(\sigma_{i_1},\sigma_{i_2}) = \sigma_{i_1}\sigma_{i_2}/\gcd(\sigma_{i_1},\sigma_{i_2}).$$

Consider an iterated algorithm for length $[\text{lcm}(\sigma_{i_1},\sigma_{i_2})\gcd(\sigma_{i_1},\sigma_{i_2})]$. It clearly has multiplicative complexity $M_{\sigma_{i_1}} M_{\sigma_{i_2}}$; therefore, we obtain

$$M_{\text{lcm}(\sigma_{i_1},\sigma_{i_2})} = M_{\sigma_{i_1}} M_{\sigma_{i_2}}/M_{\gcd(\sigma_{i_1},\sigma_{i_2})}. \tag{4.5}$$

We demonstrate some of the foregoing principles with an example.

EXAMPLE 3. Consider $GF(2)$ and $N_1 = 9$, $N_2 = 25$ ($N = 225$). We find that $\Sigma_1 = \{6, 2\}$ and $\Sigma_2 = \{20, 4\}$. Computing the least common multiples of all pairs $(\sigma_{i_1}, \sigma_{i_2})$ and attaching to these the appropriate multiplicities, we obtain the sequence, $\Sigma' = \{60, 60, 12, 12, 20, 20, 4, 4\}$; thus, $\Sigma = \Sigma_1 \cup \Sigma_2 \cup \Sigma' = \{60, 60, 12, 12, 20, 20, 4, 4, 6, 2, 20, 4\}$. Consider the aperiodic convolution for sequences of length 60 (this is lcm(20, 6)); using (4.5), we have $M_{60} = M_{20} M_6 / M_2$, and an iterated algorithm requires $M_5 M_2 M_2 M_3$ multiplications. If $M_5 = 16$, $M_3 = 6$, and $M_2 = 3$, the number is 864 multiplications over $GF(2)$. It is possible to do better than this, but this approach has the advantage of simplicity.

We are now in a position to state the principal result.

THEOREM 4.1 *Let the length of the cylic convolution over the field of constants $GF(p)$ be $N = N_1 N_2$, where $(N_1, N_2) = 1$, $(N, p) = 1$, and e_i are the orders of p mod N_i, $i = 1, 2$. Compute a cyclic convolution in accord with the method of Section 3.1. Compute an aperiodic convolution by iterating on small-degree polynomial multiplication algorithms. Then,*

$$M(N_1 N_2) = M(N_1) M(N_2), \quad if \ gcd(e_1, e_2) = 1;$$
$$M(N_1 N_2) < M(N_1) M(N_2), \quad if \ gcd(e_1, e_2) > 1.$$

PROOF. We start from (4.4) of Lemma 4.1. Using (4.5), we have

$$M(N_1 N_2) = M(N_1) + M(N_2) - 1$$
$$+ \sum_{i_1 \epsilon S_{N_1}} \sum_{i_2 \epsilon S_{N_2}} \left[\frac{\gcd(\sigma_{i_1}, \sigma_{i_2})}{M_{\gcd(\sigma_{i_1}, \sigma_{i_2})}} \right] M_{\sigma_{i_1}} M_{\sigma_{i_2}}. \quad (4.6)$$

First consider the case $\gcd(e_1, e_2) = 1$; from property $P2'$, this implies $\gcd(\sigma_{i_1}, \sigma_{i_2}) = 1$, $i_1 \epsilon S_{N_1}$, $i_2 \epsilon S_{N_2}$, for which (4.6) simplifies to

$$M(N_1 N_2) = M(N_1) + M(N_2) - 1 + [M(N_1) - 1] [M(N_2) - 1]$$
$$= M(N_1) M(N_2).$$

We have used $M_1 = 1$ and (4.1). This establishes the first result. To prove the second result, we note that since $M_l > l$, $l > 1$, the interaction term will always be strictly less than its maximum $[M(N_1) - 1][M(N_2) - 1]$ when $\gcd(e_1, e_2) > 1$. This observation immediately leads to the second result.

There are many implications of the above result; we touch on these in the following discussion.

Theorem 4.1 indicates that, over a specific field of constants $GF(p)$, the optimality of long cyclic convolutions of length $N = N_1 N_2$, $(N_1, N_2) = 1$, depends on whether the exponents e_i of p mod N_i, $i = 1, 2$, are relatively prime. This result is tied to the *tensor product* of the roots (orbits) of the irreducible polynomials over $GF(p)$ associated with the subfields of $GF(p^e)$. To see this, we work with the matrix C associated with the bilinear algorithm for length N cyclic convolution. The matrix C is $(N \times (M(N))$-dimensional and composed of submatrices C_i, $i \epsilon S_{N_1 N_2} \cup \{0\}$. Consider a typical submatrix C_i; from Theorem 3.1, the jth row of C_i may be obtained from

$$\psi_{ij} = (1/N) \sum_{m=0}^{2\sigma_i - 1} r_m \, tr(\alpha^{i(m-j)}), \qquad (4.7)$$

where α is a primitive element of $GF(p^e)$. We see that $\xi = \alpha^{(\frac{p^e-1}{N})}$ is a primitive Nth root of unity in $GF(p^e)$ and, due to the manner in which cyclotomic sets are constructed we may write $\alpha^i = \xi^k$, where k is an integer uniquely determined by i. Let $\beta = \alpha^{(\frac{p^e-1}{N_1})}$ and $\gamma = \alpha^{(\frac{p^e-1}{N_2})}$ be primitive N_1th and N_2th roots of unity in $GF(p^e)$, respectively. We then have $\alpha^i = \xi^k = \beta^{k_1}\gamma^{k_2}$, where the pair of integers k_1, k_2 is uniquely determined by k. This permits us to write (4.7) as

$$
\begin{aligned}
\psi_{ij} &= (1/N) \sum_{m=0}^{2\sigma_i - 1} r_m \, tr(\beta^{k_1(m-j)} \gamma^{k_2(m-j)}) \\
&= (1/N) \sum_{m=0}^{2\sigma_i - 1} r_m \sum_{l=0}^{\sigma_i - 1} (\beta^{k_1(m-j)} \gamma^{k_2(m-j)})^{p^l}. \qquad (4.8)
\end{aligned}
$$

Now, assume that $\gcd(e_1, e_2) = 1$; therefore, we have $(\sigma_{i_1}, \sigma_{i_2}) = 1$, $i_1 \epsilon S_{N_1} \cup \{0\}$, $i_2 \epsilon S_{N_2} \cup \{0\}$, from $P2'$. The Chinese Remainder Theorem for integers then guarantees the existence of integers a, b such that $a\sigma_{i_1} + b\sigma_{i_2} \equiv 1 \bmod \sigma_i$. Also note that $|S_{N_1 N_2}| = |S_{N_1}| |S_{N_2}|$ and that we can always find a σ_{i_1} and a σ_{i_2}, $i_1 \epsilon S_{N_1} \cup \{0\}$, $i_2 \epsilon S_{N_2} \cup \{0\}$, such that $\sigma_i = \sigma_{i_1}\sigma_{i_2}$, $i \epsilon S_{N_1 N_2} \cup \{0\}$. Consider the one-to-one mapping in (4.8) for which $l \to (l_1, l_2)$, where $l \equiv l_1 \bmod \sigma_{i_1}$, $l \equiv l_2 \bmod \sigma_{i_2}$. With some manipulation, we find that (4.8) takes the form

$$\psi_{ij} = (1/N) \sum_{m=0}^{2\sigma_i-1} r_m \sum_{l_1=0}^{\sigma_{i_1}-1} \sum_{l_2=0}^{\sigma_{i_2}-1} \left(\beta^{k_1(m-j)}\right)^{p^{l_1}} \left(\gamma^{k_2(m-j)}\right)^{p^{l_2}}$$

$$= (1/N) \sum_{m=0}^{2\sigma_i-1} r_m \, tr_{GF(p)}^{GF(p^{\sigma_{i_1}})} \left(\beta^{k_1(m-j)}\right) tr_{GF(p)}^{GF(p^{\sigma_{i_2}})} \left(\gamma^{k_2(m-j)}\right),$$

$$\gcd(e_1, e_2) = 1, \qquad (4.9)$$

where, for clarity, the subfield-to-prime subfield mapping for the trace operator is indicated. Since $\sigma_{i_1} = e_1$, for at least one $i_1 \epsilon S_{N_1}$, and $\sigma_{i_2} = e_2$, for at least one $i_2 \epsilon S_{N_2}$, by $P3'$, the finite fields of largest cardinality used in the mapping down to $GF(p)$ are $GF(p^{e_1})$ and $GF(p^{e_2})$. The inclusion relationships are as depicted below:

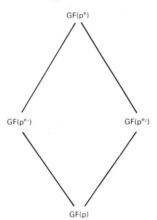

We see from (4.9) that every Nth root of unity associated with a cyclotomic set of cardinality σ_i is formed as the tensor product of N_1th and N_2th roots of unity associated with cyclotomic sets of cardinalities σ_{i_1} and σ_{i_2}, respectively. Furthermore, a representation of the form (4.3) is valid for $\gcd(e_1, e_2) = 1$.

The situation for $\gcd(e_1, e_2) > 1$ is now easily seen from the inclusion relationships depicted below in which $s = \gcd(e_1, e_2)$:

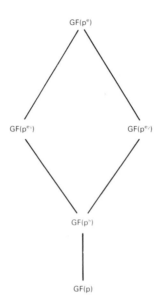

The mappings are now from $GF(p^{e_1})$ and $GF(p^{e_2})$ to $GF(p^s)$, followed by a mapping from $GF(p^s)$ to $GF(p)$. The result is a representation more complex than (4.3) and a multiplicative complexity lower than $M(N_1)M(N_2)$ when $\gcd(e_1, e_2) > 1$. Tables 4.1 and 4.2 provide a comparison of the multiplicative complexities $M(N_1 N_2)$ and $M(N_1)M(N_2)$ for moderate-to-long length cyclic convolution over the fields of constants $GF(2)$ and $GF(3)$, respectively.

4.1.2 Some Specific Results Pertinent to Long Length Design

In this section, a number of pertinent results (in the sense of commonly used selections for the factors N_1, N_2) are presented relative to long length cyclic convolution design. The interested reader may find more details in [4.6]. The results all follow from Lemma 4.1, Theorem 4.1, and the properties previously presented.

COROLLARY 4.1 *Let the factor lengths for cyclic convolution over* $GF(p)$ *be expressible as* $N_1 = p_1^{n_1}$ *and* $N_2 = p_2^{n_2}$, *where* p_1 *and* p_2 *are primes* $(\neq p)$. *The order of* $p \bmod p_i$ *is denoted as* e'_i, $i = 1, 2$. *If* $\gcd(p_i^{n_i-1}, e'_j) = 1$, $i, j = 1, 2; i \neq j$, *then*

Table 4.1: Comparison of the Multiplicative Complexities $M(N)$ and $M(N_1)M(N_2)$ Over the Field of Constants $GF(2)$ $(N = N_1 N_2)$.

N	N_1	N_2	e_1	e_2	$M(N_1)$	$M(N_2)$	$M(N_1)M(N_2)$	$M(N)$	$M(N)/M(N_1)M(N_2)$	$M(N)/N$
15	5	3	4	2	10	4	40	31	.7778	2.0667
33	11	3	10	2	49	4	196	148	.7551	4.4848
35	7	5	3	4	13	10	130	130	1.0	3.7143
51	17	3	8	2	55	4	220	166	.7545	3.2549
55	11	5	10	4	49	10	490	346	.7061	6.2909
85	17	5	8	4	55	10	550	280	.5091	3.2941
91	13	7	12	3	55	13	715	391	.5469	4.2967
93	31	3	5	2	97	4	388	388	1.0	4.1720
117	13	9	12	6	55	22	1210	508	.4198	4.3419
205	41	5	20	4	289	10	2890	1450	.5017	7.0732
315	63	5	6	4	178	10	1780	1285	.7219	4.0794
455	65	7	12	3	280	13	3640	2020	.5550	4.4396
511	73	7	9	3	289	13	3757	2029	.5401	3.9706
663	221	3	24	2	1405	4	5620	4216	.7502	6.3590
765	85	9	8	6	280	22	6160	4207	.6830	5.4993
949	73	13	9	12	289	55	15895	8119	.5108	8.5553
1989	117	17	12	8	500	55	27940	12982	.4646	6.5269
6643	949	7	36	3	8119	13	105547	56839	.5385	8.5562

Table 4.2: Comparison of the Multiplicative Complexities $M(N)$ and $M(N_1)M(N_2)$ Over the Field of Constants $GF(3)$ $(N = N_1N_2)$.

N	N_1	N_2	e_1	e_2	$M(N_1)$	$M(N_2)$	$M(N_1)M(N_2)$	$M(N)$	$M(N)/M(N_1)M(N_2)$	$M(N)/N$
10	5	2	4	1	10	2	20	20	1.0	2.0000
20	5	4	4	2	10	5	50	41	.8200	2.0500
28	7	4	6	2	19	5	95	77	.8105	2.7500
34	17	2	16	1	82	2	164	164	1.0	4.8235
35	7	5	6	4	19	10	190	136	.7158	3.8857
40	8	5	2	4	11	10	110	83	.7545	2.0750
44	11	4	5	2	33	5	165	165	1.0	3.7500
55	11	5	5	4	33	10	330	330	1.0	6.0000
56	8	7	2	6	11	19	209	155	.7416	2.7679
68	17	4	16	2	82	5	410	329	.8024	4.8382
85	17	5	16	4	82	10	820	415	.5061	4.8824
91	13	7	3	6	25	19	475	259	.5453	2.8462
205	41	5	8	4	136	10	1360	685	.5037	3.3415
455	91	5	6	4	259	10	2590	1888	.7290	4.1495
656	41	16	8	4	136	29	3944	2189	.5550	3.3369
697	41	17	8	16	136	82	11152	3457	.3100	4.9598
5299	757	7	9	6	3025	19	57475	30259	.5265	5.7103
6056	757	8	9	2	3025	11	33275	33275	1.0	5.4946

$$M(N_1N_2) = M(N_1)M(N_2) - \left[1 - \frac{gcd(e_1', e_2')}{M_{gcd(e_1', e_2')}}\right][M(N_1)-1][M(N_2)-1].$$
(4.10)

PROOF. From *P6′*, we may write σ_{i_1}, $i_1 \epsilon S_{N_1}$, and σ_{i_2}, $i_2 \epsilon S_{N_2}$, in the form

$$\sigma_{i_1} = e_1' p_1^\ell, \ 0 \le \ell \le n_1 - 1; \ \sigma_{i_2} = e_2' p_2^k, \ 0 \le k \le n_2 - 1.$$

The integers ℓ and k depend on i_1 and i_2, respectively. Note that $gcd(\sigma_{i_1}, \sigma_{i_2}) = gcd(e_1' p_1^\ell, e_2' p_2^k) = gcd(e_1', e_2')$. Inserting this into (4.6), we obtain

$$
\begin{aligned}
M(N_1N_2) \ &= \ M(N_1) + M(N_2) - 1 \\
&+ \sum_{i_1 \epsilon S_{N_1}} \sum_{i_2 \epsilon S_{N_2}} \left[\frac{gcd(e_1', e_2')}{M_{gcd(e_1', e_2')}}\right] M_{\sigma_{i_1}} M_{\sigma_{i_2}} \\
&= \ M(N_1) + M(N_2) - 1 \\
&+ \left[\frac{gcd(e_1', e_2')}{M_{gcd(e_1', e_2')}}\right] \sum_{i_1 \epsilon S_{N_1}} M_{\sigma_{i_1}} \sum_{i_2 \epsilon S_{N_2}} M_{\sigma_{i_2}},
\end{aligned}
$$

which, after use of (4.1), provides the desired result.

As an example of the application of Corollary 4.1, take the factors $N_1 = 3^2, N_2 = 5^2$; $N = N_1N_2 = 225$. Notice that for moderate-to-large N_1N_2, the following approximation obtained from (4.10) is accurate:

$$M(N_1N_2) \cong \left[\frac{gcd(e_1', e_2')}{M_{gcd(e_1', e_2')}}\right] M(N_1)M(N_2).$$

Since $e_1' = 2$, $e_2' = 4$ over $GF(2)$, the factor in brackets has the value $2/3$; thus, $M(N_1N_2) \cong 0.67M(N_1)M(N_2)$.

COROLLARY 4.2 *Let the factor lengths for cyclic convolution over $GF(p)$ be expressible as $N_1 = p_1^{n_1}$ and $N_2 = p_2^{n_2}$, where p_1 and p_2*

are primes ($\neq p$). The order of $p \bmod p_i$ is denoted as e_i', $i = 1, 2$. If $p_1^{n_1-1} | e_2'$ and $p_2^{n_2-1} | e_1'$, then

$$
\begin{aligned}
M(N_1 N_2) \;=\; & M(N_1) M(N_2) - \Big\{ [M(N_1) - 1][M(N_2) - 1] \\
& - (N_1 - 1)(N_2 - 1) \left[\frac{M_{lcm(e_1', e_2')}}{lcm(e_1', e_2')} \right] \Big\} .
\end{aligned}
\tag{4.11}
$$

PROOF. We write σ_{i_1}, $i_1 \epsilon S_{N_1}$, and σ_{i_2}, $i_2 \epsilon S_{N_2}$, in the form indicated by $P6'$. Recall that $\gcd(\sigma_{i_1}, \sigma_{i_2}) = \sigma_{i_1} \sigma_{i_2} / \mathrm{lcm}(\sigma_{i_1}, \sigma_{i_2})$; to evaluate $\gcd(\sigma_{i_1}, \sigma_{i_2})$, we first evaluate $\mathrm{lcm}(\sigma_{i_1}, \sigma_{i_2})$. We have, for, $0 \leq \ell \leq n_1 - 1;\ 0 \leq k \leq n_2 - 1$,

$$
\mathrm{lcm}(\sigma_{i_1}, \sigma_{i_2}) = \mathrm{lcm}(e_1' p_1^\ell, e_2' p_2^k) = \mathrm{lcm}(e_1', e_2');
$$

thus,

$$
\gcd(\sigma_{i_1}, \sigma_{i_2}) = \sigma_{i_1} \sigma_{i_2} / \mathrm{lcm}(e_1', e_2').
$$

Using an iterated polynomial multiplication algorithm, we obtain

$$
M_{\gcd(\sigma_{i_1}, \sigma_{i_2})} = M_{\sigma_{i_1}} M_{\sigma_{i_2}} / M_{\mathrm{lcm}(e_1', e_2')}.
$$

Substituting the expressions for $\gcd(\sigma_{i_1}, \sigma_{i_2})$ and $M_{\gcd(\sigma_{i_1}, \sigma_{i_2})}$ into (4.6), leads to

$$
\begin{aligned}
M(N_1 N_2) \;=\; & M(N_1) + M(N_2) - 1 \\
& + \sum_{i_1 \epsilon S_{N_1}} \sum_{i_2 \epsilon S_{N_2}} \left[\frac{M_{\mathrm{lcm}(e_1', e_2')}}{\mathrm{lcm}(e_1', e_2')} \right] \sigma_{i_1} \sigma_{i_2} \\
\;=\; & M(N_1) + M(N_2) - 1 \\
& + \left[\frac{M_{\mathrm{lcm}(e_1', e_2')}}{\mathrm{lcm}(e_1', e_2')} \right] \sum_{i_1 \epsilon S_{N_1}} \sum_{i_2 \epsilon S_{N_2}} \sigma_{i_1} \sigma_{i_2},
\end{aligned}
$$

which, after application of $P5'$, provides the desired result.

The result of Corollary 4.2 is not as useful for design purposes as that of Corollary 4.1; however, we do note that if N_1 and N_2 are both primes, $\gcd(e_1, e_2) = 1$, and both corollaries give the same result for $M(N_1 N_2)$. From (4.10) or (4.2), the result is

$$M(N_1 N_2) \quad = \quad \left[(N_1 - 1)\frac{M_{e_1}}{e_1} + 1 \right] \left[(N_2 - 1)\frac{M_{e_2}}{e_2} + 1 \right],$$
$$(N_1, N_2 \ prime). \tag{4.12}$$

4.2 Multiplicative Complexity

For length N cyclic convolution over the field of constants $GF(p)$, the multiplicative complexity of the algorithm presented in Section 3.1 is given by $M(N) = 1 + \sum_{i \in S_N} M_{\sigma_i}$. We know that over any field $\mathcal{F}$, the minimum of $M(N)$, which we denote by μ, satisfies $\mu > 2N - |S_N| - 1$. If the field is sufficiently large, ie., $|\mathcal{F}| > 2N - 2$, then we have that $\mu = 2N - |S_N| - 1$. The latter case follows from the fact that a sufficient number of distinct constants exist and the Toom-Cook procedure can be employed to achieve the lower bound. In either case, we have seen that the cardinality of the index set, $|S_N|$, plays an important role in establishing a lower bound on multiplicative complexity. For a cyclic convolution of length N, $(N, p) = 1$ and $N|(p^e - 1)$, each cyclotomic set corresponds to an irreducible factor of the polynomial $u^N - 1$ over $GF(p)$ and $|S_N| + 1$ is the total number of such factors. Finding these factors (and even the *number* of factors) is a difficult task; an approach was suggested in Section 3.1.5 whereby the cyclotomic polynomial factors were first found over the field of complex numbers, and then each factor was examined to determine its irreducibility over $GF(p)$.

Polynomial theory and the problem of factoring polynomials have a long and fascinating history. The reader may refer to [4.7] or [4.8] for an excellent exposition of this work. In 1967, E. Berlekamp [4.9, 4.10] devised an ingenious algorithm which factors degree N univariate polynomials over $GF(p)$. The running time (computational complexity) of this method is $\mathcal{O}(N^3 + pLN^2)$, where L is the actual number of factors over $GF(p)$. Following the idea of Berlekamp, random choices were introduced in [4.11] into the algorithm for factorization over large finite fields; this led to an *expected* computational complexity of at most $\mathcal{O}[N^3 (\log N)^3 \log p]$. Further improvements to Berlekamp's algorithm using probabilistic methods are discussed in [4.8]. A goal of much of the work has been to reduce a complexity which is exponential in N to one polynomial in $N \log p$. No closed form expression for the number of irreducible factors over $GF(p)$ of a large degree polynomial is known; however, it is known that an Nth degree polynomial has an *average* of $\log N$ factors as p tends to infinity [4.7]. We see, therefore, that it is difficult to determine $|S_N|$ over $GF(p)$ when N is arbitrarily large. To obtain a rough idea of $|S_N|$, we have developed a simple program to compute a lower bound for the multiplicative complexity $M(N)$ of sufficiently large length N cyclic convolution over $GF(p)$. The program is given in Appendix B.

It appears that an *estimate* of $|S_N|$ may be obtained as

$$|S_N| \cong m(\frac{2p-1}{p-1}),$$

where m is the solution to

$$p^m = (\frac{p}{2p-1})[(N - m + 1)\log_e p + 1].$$

For example, when $N = 63$ and the field of constants is $GF(2)$, we have $m \cong 4.8083$ and $|S_N| \cong 14.4250$; we find the exact figure to be $|S_N| = 12$. Work with the program also indicates that the minimum number of multiplications is closer to $3N - |S_N|$ over $GF(2)$, as opposed to $2N - |S_N|$. This type of behavior was observed in [4.12] for the situation where $|S_N| = 1$ (multiplication of polynomials each of degree $N - 1$, reduced modulo a monic, irreducible polynomial of degree N). In the next section, we take a brief look at the distance properties of linear codes generated by embedding a particular system of bilinear forms in a cyclic convolution.

4.3   A Distance Bound

Let Ψ be the system of bilinear forms given by $\psi = X\mathbf{y}$, where the $(k \times d)$-dimensional matrix X is defined as in (2.15). We have seen that we may embed this system in the larger N-dimensional circulant system of (2.19), where $N = k + d - 1$. We have obtained a bilinear algorithm of the form $C(A\mathbf{x} \bullet B\mathbf{y})$ requiring $M(N)$ multiplications for the larger system. It is obvious that with proper planning, we can drop $N - k$ consecutive rows of C to obtain an algorithm for computing the original system Ψ. Call the resulting $(k \times M(N))$-dimensional matrix C (to avoid further notation); we have

$$\psi = X\mathbf{y} = C(A\mathbf{x} \bullet B\mathbf{y}), \tag{4.13}$$

where we may think of C as the generator matrix of an (n, k, d) linear code, $M(N) = n$, with symbols from $GF(p)$. Note that X and C have k $\ell.i.$ rows; thus, C has k $\ell.i.$ columns. Dropping $N - k$ rows from the original matrix C has reduced the number of $\ell.i.$ columns. The original matrix C had N $\ell.i.$ columns, easily seen from the fact that there were M_{σ_i} columns in the block C_i of C, $i \epsilon S_N$, of which σ_i were $\ell.i.$ Using $P5'$, we see that C had a total of $\sum_{i \epsilon S_N \cup \{0\}} \sigma_i = N$ $\ell.i.$ columns.

Recall that the linear code is a vector subspace of $GF^n(p)$. Since elementary row operations on C leave the code space invariant, we may also choose a basis of the subspace such that C is in *systematic* form. This is equivalent to making the $\ell.i.$ columns of C the first k columns, followed by multiplication from the left with a nonsingular $(k \times k)$-dimensional matrix E having entries from $GF(p)$, i.e.,

$$E\psi = EX\mathbf{y} = [I_k \mid C']\mathbf{m}. \tag{4.14}$$

The $(k \times (n - k))$-dimensional matrix C' results as the product of E with the submatrix consisting of the $n - k$ linearly dependent columns of C. The right-hand side (r.h.s.) of (4.14) is a bilinear algorithm for computing the system $E\psi$. Linear codes generated by C and EC have precisely the same distance structure.

Now, let E_ℓ denote the matrix consisting of the first ℓ rows of E, $\ell = 1, 2, \ldots, k$. From (4.14), we have

$$E_\ell\psi = E_\ell X\mathbf{y} = [I_\ell | O_{k-\ell} | C'_\ell]\mathbf{m}, \tag{4.15}$$

where C'_ℓ is the matrix consisting of the first ℓ rows of C', $\ell = 1, 2, \ldots, k$. We now have a bilinear algorithm for computing $E_\ell X\mathbf{y}$ requiring $n-k+\ell$ multiplications. Accordingly, we drop the elements (multiplications) $m_{\ell+1}, m_{\ell+2}, \ldots, m_k$ from $\mathbf{m}$ and call the resulting vector $\mathbf{m}_\ell$; this leads to the expression

$$E_\ell\psi = E_\ell X\mathbf{y} = [I_\ell | C'_\ell]\mathbf{m}_\ell. \tag{4.16}$$

In coding theory terms, we have both *punctured* and *shortened* (in regard to the dimension) the original code by $k - \ell$ symbols. The punctured symbols, however, have been carefully chosen so that the

minimum distance does not change. In the sequel, an example will be
provided to illustrate the operations.

Let $C_\ell = [I_\ell | C_\ell']$; a typical codeword is given by $\mathbf{c}^T = \mathbf{a}^T C_\ell$, where
$\mathbf{a} \in GF^\ell(p)$, $\mathbf{a} \neq \mathbf{0}$, is the vector of information symbols. Consider the
bilinear form $\mathbf{a}^T E_\ell X \mathbf{y} = \mathbf{a}^T C_\ell \mathbf{m}_\ell$. As before, we denote the Hamming
weight of a vector $\mathbf{v}$ with entries from $GF(p)$ as $w(\mathbf{v})$. Since $\mathbf{a}^T E_\ell X$
has d $\ell.i.$ columns over $GF(p)$ in the indeterminates $x_0, x_1, \ldots, x_{k-1}$,
we have

$$\sum_{\mathbf{a} \in GF^\ell(p)} w(\mathbf{c}) \geq d(p^\ell - 1), \tag{4.17}$$

where $d = N - \ell + 1$. On the other hand, using standard arguments,
we find that each coordinate position of $\mathbf{c}$ contains at most $(p-1)p^{\ell-1}$
nonzero field elements. Since there are $n - k + \ell$ such positions, we
obtain

$$\sum_{\mathbf{a} \in GF^\ell(p)} w(\mathbf{c}) \leq (n - k + \ell)(p-1)p^{\ell-1}. \tag{4.18}$$

The inequalities (4.17) and (4.18) lead to

$$n \geq k + d(\frac{p}{p-1}) - \left[\ell + d(\frac{p}{p-1})p^{-\ell}\right]. \tag{4.19}$$

The quantity in brackets is minimum (resulting in the glb) for a value
of ℓ equal to ℓ^*, where

$$\ell^* = \left\lceil \log_p \left[d(\frac{p}{p-1})/\log_p e \right] \right\rceil; \tag{4.20}$$

thus, the lower bound becomes

$$n \geq k + d(\frac{p}{p-1}) - \left\{ \left\lceil \log_p \left[d(\frac{p}{p-1})/\log_p e \right] \right\rceil + \log_p e \right\}. \tag{4.21}$$

We see that the lower bound is asymptotically $k + d + \mathcal{O}(d)$ as
the field size increases and is maximal for GF(2), being asymptotically
$k + 2d + \mathcal{O}(d)$. It is interesting to observe that (4.21) implies that the
number of parity-check symbols, $n - k$, required to achieve minimum
weight d is bounded as

$$n - k \geq d(\frac{p}{p-1}) - \left\{ \left\lceil \log_p \left[d(\frac{p}{p-1})/\log_p e \right] \right\rceil + \log_p e \right\},$$

which is essentially the Plotkin bound [4.13]. Of interest also is the weight distribution of codes constructed in the above fashion. The following example illustrates some of the described principles.

EXAMPLE 4. Consider the case of cyclic convolution of length $N = k + d - 1 = 5$ over $GF(2)$. The bilinear algorithm is of the form $C(A\mathbf{x} \bullet B\mathbf{y})$, where $C = [C_0|C_3]$. Assume $k = 3$ and $d = 3$; dropping $N - k = 2$ rows of C results in

$$C = \begin{bmatrix} 1 & 1 & 1 & 1 & 1 & 0 & 1 & 1 & 1 & 1 \\ 1 & 1 & 0 & 1 & 1 & 1 & 0 & 1 & 0 & 0 \\ 1 & 1 & 1 & 1 & 0 & 1 & 0 & 0 & 0 & 1 \end{bmatrix}.$$

Evidently, $M(N) = n = 10$ multiplications are required. Define the weight enumerator polynomial as $A(u) = \sum_{i=0}^{n} A_i u^i$, where the code contains A_i codewords of Hamming weight i. The weight enumerator for the code under consideration is $A(u) = 1 + u^4 + 3u^5 + 2u^6 + u^9$, and C is the generator matrix of a $(10, 3, 4)$ linear code over $GF(2)$.

In this example, locating the $k = 3$ linear independent columns of C is simple; we find that one possible arrangement of columns gives

$$E X \mathbf{y} = [I_3|C']\mathbf{m},$$

where

$$E = \begin{bmatrix} 1 & 1 & 1 \\ 1 & 1 & 0 \\ 1 & 0 & 1 \end{bmatrix} \text{ and } C' = \begin{bmatrix} 1 & 1 & 0 & 1 & 0 & 1 & 0 \\ 0 & 0 & 1 & 1 & 0 & 1 & 1 \\ 0 & 0 & 1 & 1 & 1 & 1 & 0 \end{bmatrix}.$$

Now let $\ell = 2$, the result of which is to puncture the third symbol of the code and shorten the generator matrix by one row. The proper expression is

$$E_2 X \mathbf{y} = C_2 \mathbf{m}_2 = [I_2|C_2']\mathbf{m}_2.$$

The weight enumerator associated with C_2 is $A(u) = 1 + 2u^5 + u^6$; the code is a $(9, 2, 5)$ code. When $\ell = k = 3$, we have $\sum_{\mathbf{a} \in GF^\ell(p)} w(\mathbf{c}) =$

$40 > d(p^\ell - 1) = 28$ and for $\ell = 2$, $\sum_{\mathbf{a}\epsilon GF^\ell(p)} w(\mathbf{c}) = 16 > d(p^\ell - 1) = 15$. In both cases, the sum of Hamming weights is also seen to satisfy (4.18) with equality.

4.4   References

[**4.1**] R. C. Agarwal and C. S. Burrus, "Fast One Dimensional Convolution by Multidimensional Techniques," *IEEE Trans. Acoust., Speech, Signal Processing*, vol. ASSP-22, pp. 1-10, Feb. 1974.

[**4.2**] R. C. Agarwal and J. W. Cooley, "New Algorithms for Digital Convolution," *IEEE Trans. Acoust., Speech, Signal Proceesing*, vol. ASSP-25, pp. 392-410, Oct. 1977.

[**4.3**] I. J. Good, "The Interaction Algorithm and Practical Fourier Analysis," *Journ. Royal Statist. Soc.*, ser. B, vol. 20, pp. 361-372, 1958; addendum, vol. 22, pp. 372-375, 1960.

[**4.4**] D. P. Kolba and T. W. Parks, "A Prime Factor FFT Algorithm Using High-Speed Convolution," *IEEE Trans. Acoust., Speech, Signal Processing*, vol. ASSP-25, pp. 90-103, Aug. 1977.

[**4.5**] S. Winograd, "On Computing the Discrete Fourier Transform," *Math. of Computat.*, vol. 32, no. 141, pp. 175-199, Jan. 1978.

[**4.6**] M. D. Wagh and S. D. Morgera, "Cyclic Convolution Algorithms over Finite Fields: Multidimensional Considerations," in Proc. *IEEE Int. Conf. Acoust., Speech, Signal Processing*, Atlanta, GA, pp. 327-330, Mar. 1981.

[**4.7**] D. E. Knuth, *The Art of Computer Programming, Vol. 2, Seminumerical Algorithms*, Addison Wesley, 1981.

[**4.8**] E. Kaltofen, "Factorization of Polynomials," *Computer Algebra, Symbolic and Algebraic Computation*, Eds. B. Buchberger, et al, Springer-Verlag, 1983.

[**4.9**] E. R. Berlekamp, "Factoring Polynomials over Large Finite Fields," *Math. of Comput.*, vol. 24, pp. 713-735, 1970.

[**4.10**] E. R. Berlekamp, *Algebraic Coding Theory*, McGraw-Hill, 1968.

Finite Fields," *SIAM*

"On the Complexity
Comp. Sci., vol. 22,

or-Correcting Codes,

4.5   Appendix A

Proofs of Properties $P1'$

Property $P1'$ follows by cons...
treated in Appendix A of Chap...
the proof of property $P3'$.

PROOF ($P3'$).   The expone...
$N|(p^e - 1)$; also, σ_i is the small...
$i = j(p^e - 1)/N, \; j\epsilon[1, N-1]$...
$1)/N]$ or $N|j(p^n - 1)$. When j
smallest solution $n = e$; there...
$1)/N, \; \sigma_i = e$.

PROOF ($P4'$).   If N is prime,...
factor. Then, $N|j(p^n - 1)$ resul...
$P3'$, the smallest solution is n
$i, \; \sigma_i = e$.

PROOF ($P6'$).   By induction...
$N = q^n$ is a divisor of $e'q^{n-1}$...
$n = 1$. The expression $q^n|(p^e$...
$p^{qe_n} = (1 + kq^n)^q = 1 + k'q^{n+1}$,...
expression $q^{n+1}|(p^{qe_n} - 1)$, whi...
exponent of q^{n+1}. Moreover, q^n...
or $e_n|e_{n+1}$. These relations sho...

PROOF ($P7'$).   For every...
$1)$, or $(p^e - 1)|k(p^{\sigma_i} - 1)$. Recall...
$N-1]$; hence, $N|j(p^{\sigma_i} - 1)$. For...
satisfy $N|j(p^{\sigma_i} - 1)$. In a similar...
$j_2 \leq N_2 - 1$, can be partitione...
$N = N_1 N_2$ and $(N_1, N_2) = 1$, th...
as

$$j \equiv j_2 N_1$$

[**4.11**] M. O. Rabin, "Probabilistic Algorithms in Finite Fields," *SIAM Journ. Comp.*, vol. 9, pp. 273-280, 1980.

[**4.12**] A. Lempel, G. Seroussi, and S. Winograd, "On the Complexity of Multiplication in Finite Fields," *Theo. Comp. Sci.*, vol. 22, pp. 285-296, 1983.

[**4.13**] W. W. Peterson and E.J. Weldon, Jr., *Error-Correcting Codes*, MIT Press, 1972.

4.5   Appendix A

Proofs of Properties $P1' - P7'$

Property $P1'$ follows by construction of the cyclotomic sets, $P2'$ is treated in Appendix A of Chapter 3, and $P5'$ is obvious. We start with the proof of property $P3'$.

PROOF ($P3'$).   The exponent e is the smallest integer such that $N|(p^e - 1)$; also, σ_i is the smallest n such that $(p^e - 1)|i(p^n - 1)$. Since $i = j(p^e - 1)/N$, $j\epsilon[1, N - 1]$, this implies $(p^e - 1)|[j(p^n - 1)(p^e - 1)/N]$ or $N|j(p^n - 1)$. When $j = 1$, by definition, $N|(p^n - 1)$ has the smallest solution $n = e$; therefore, for the corresponding $i = (p^e - 1)/N$, $\sigma_i = e$.

PROOF ($P4'$).   If N is prime, N and j, $j\epsilon[1, N - 1]$, have no common factor. Then, $N|j(p^n - 1)$ results in $N|(p^n - 1)$ and, as in the proof of $P3'$, the smallest solution is $n = e$. For N a prime, therefore, for all i, $\sigma_i = e$.

PROOF ($P6'$).   By induction, it is shown that the exponent e_n of $N = q^n$ is a divisor of $e'q^{n-1}$. The statement is obviously true for $n = 1$. The expression $q^n|(p^{e_n} - 1)$ results, for some integer k, in $p^{qe_n} = (1 + kq^n)^q = 1 + k'q^{n+1}$, where k' is an integer. This leads to the expression $q^{n+1}|(p^{qe_n} - 1)$, which implies $e_{n+1}|qe_n$, where e_{n+1} is the exponent of q^{n+1}. Moreover, $q^{n+1}|(p^{e_{n+1}} - 1)$ implies that $q^n|(p^{e_{n+1}} - 1)$ or $e_n|e_{n+1}$. These relations show that $e_{n+1} = e_n$ or qe_n, as required.

PROOF ($P7'$).   For every member k of S_i, $kp^{\sigma_i} \equiv k \bmod(p^e - 1)$, or $(p^e - 1)|k(p^{\sigma_i} - 1)$. Recall that k is of the form $j(p^e - 1)/N$, $j\epsilon[1, N-1]$; hence, $N|j(p^{\sigma_i} - 1)$. For every $\sigma_i \epsilon \sum$, σ_i values of j, $j\epsilon[1, N-1]$, satisfy $N|j(p^{\sigma_i} - 1)$. In a similar fashion, the sets $1 \le j_1 \le N_1 - 1$, $1 \le j_2 \le N_2 - 1$, can be partitioned using $\sum_1$ and $\sum_2$, respectively. If $N = N_1N_2$ and $(N_1, N_2) = 1$, then any $j\epsilon[1, N - 1]$ can be represented as

$$j \equiv j_2N_1 + j_1N_2 \bmod N.$$

$$n - k \geq d(\frac{p}{p-1}) - \left\{ \left\lceil \log_p \left[d(\frac{p}{p-1})/\log_p e \right] \right\rceil + \log_p e \right\},$$

which is essentially the Plotkin bound [4.13]. Of interest also is the weight distribution of codes constructed in the above fashion. The following example illustrates some of the described principles.

EXAMPLE 4. Consider the case of cyclic convolution of length $N = k + d - 1 = 5$ over $GF(2)$. The bilinear algorithm is of the form $C(\mathbf{Ax} \bullet \mathbf{By})$, where $C = [C_0|C_3]$. Assume $k = 3$ and $d = 3$; dropping $N - k = 2$ rows of C results in

$$C = \begin{bmatrix} 1 & 1 & 1 & 1 & 1 & 0 & 1 & 1 & 1 & 1 \\ 1 & 1 & 0 & 1 & 1 & 1 & 0 & 1 & 0 & 0 \\ 1 & 1 & 1 & 1 & 0 & 1 & 0 & 0 & 0 & 1 \end{bmatrix}.$$

Evidently, $M(N) = n = 10$ multiplications are required. Define the weight enumerator polynomial as $A(u) = \sum_{i=0}^{n} A_i u^i$, where the code contains A_i codewords of Hamming weight i. The weight enumerator for the code under consideration is $A(u) = 1 + u^4 + 3u^5 + 2u^6 + u^9$, and C is the generator matrix of a $(10, 3, 4)$ linear code over $GF(2)$.

In this example, locating the $k = 3$ linear independent columns of C is simple; we find that one possible arrangement of columns gives

$$EX\mathbf{y} = [I_3|C']\mathbf{m},$$

where

$$E = \begin{bmatrix} 1 & 1 & 1 \\ 1 & 1 & 0 \\ 1 & 0 & 1 \end{bmatrix} \text{ and } C' = \begin{bmatrix} 1 & 1 & 0 & 1 & 0 & 1 & 0 \\ 0 & 0 & 1 & 1 & 0 & 1 & 1 \\ 0 & 0 & 1 & 1 & 1 & 1 & 0 \end{bmatrix}.$$

Now let $\ell = 2$, the result of which is to puncture the third symbol of the code and shorten the generator matrix by one row. The proper expression is

$$E_2 X\mathbf{y} = C_2 \mathbf{m}_2 = [I_2|C_2']\mathbf{m}_2.$$

The weight enumerator associated with C_2 is $A(u) = 1 + 2u^5 + u^6$; the code is a $(9, 2, 5)$ code. When $\ell = k = 3$, we have $\sum_{\mathbf{a} \in GF^{\ell}(p)} w(\mathbf{c}) =$

$40 > d(p^\ell - 1) = 28$ and for $\ell = 2$, $\sum_{\mathbf{a} \in GF^\ell(p)} w(\mathbf{c}) = 16 > d(p^\ell - 1) = 15$. In both cases, the sum of Hamming weights is also seen to satisfy (4.18) with equality.

4.4 References

[**4.1**] R. C. Agarwal and C. S. Burrus, "Fast One Dimensional Convolution by Multidimensional Techniques," *IEEE Trans. Acoust., Speech, Signal Processing*, vol. ASSP-22, pp. 1-10, Feb. 1974.

[**4.2**] R. C. Agarwal and J. W. Cooley, "New Algorithms for Digital Convolution," *IEEE Trans. Acoust., Speech, Signal Proceesing*, vol. ASSP-25, pp. 392-410, Oct. 1977.

[**4.3**] I. J. Good, "The Interaction Algorithm and Practical Fourier Analysis," *Journ. Royal Statist. Soc.*, ser. B, vol. 20, pp. 361-372, 1958; addendum, vol. 22, pp. 372-375, 1960.

[**4.4**] D. P. Kolba and T. W. Parks, "A Prime Factor FFT Algorithm Using High-Speed Convolution," *IEEE Trans. Acoust., Speech, Signal Processing*, vol. ASSP-25, pp. 90-103, Aug. 1977.

[**4.5**] S. Winograd, "On Computing the Discrete Fourier Transform," *Math. of Computat.*, vol. 32, no. 141, pp. 175-199, Jan. 1978.

[**4.6**] M. D. Wagh and S. D. Morgera, "Cyclic Convolution Algorithms over Finite Fields: Multidimensional Considerations," in Proc. *IEEE Int. Conf. Acoust., Speech, Signal Processing*, Atlanta, GA, pp. 327-330, Mar. 1981.

[**4.7**] D. E. Knuth, *The Art of Computer Programming, Vol. 2, Seminumerical Algorithms*, Addison Wesley, 1981.

[**4.8**] E. Kaltofen, "Factorization of Polynomials," *Computer Algebra, Symbolic and Algebraic Computation*, Eds. B. Buchberger, et al, Springer-Verlag, 1983.

[**4.9**] E. R. Berlekamp, "Factoring Polynomials over Large Finite Fields," *Math. of Comput.*, vol. 24, pp. 713-735, 1970.

[**4.10**] E. R. Berlekamp, *Algebraic Coding Theory*, McGraw-Hill, 1968.

Substituting this expression in $N|j(p^{\sigma_i} - 1)$, it is possible to conclude that the sum of the fractions, $j_2(p^{\sigma_i} - 1)/N_2 + j_1(p^{\sigma_i} - 1)/N_1$, is an integer. Since the denominators are relatively prime, each fraction itself is an integer. The expression $N|j(p^{\sigma_i} - 1)$ is, therefore, equivalent to $N_1|j_1(p^{\sigma_i} - 1)$ and $N_2|j_2(p^{\sigma_i} - 1)$. For every $\sigma_{i_1} \in \sum_1$ and $\sigma_{i_2} \in \sum_2$, there are σ_{i_1} values of j_1 for $1 \le j_1 \le N - 1$ satisfying $N_1|j_1(p^{\sigma_{i_1}} - 1)$ and σ_{i_2} values of j_2 for $1 \le j_2 \le N - 1$ satisfying $N_2|j_2(p^{\sigma_{i_2}} - 1)$. This implies that there are $\sigma_{i_1}\sigma_{i_2}$ values of j such that $\sigma_i = \mathrm{lcm}(\sigma_{i_1}, \sigma_{i_2})$. These values are grouped together in sets of cardinality σ_i. There are, therefore, $\sigma_{i_1}\sigma_{i_2}/\sigma_i = \gcd(\sigma_{i_1}, \sigma_{i_2})$ sets of cardinality $\sigma_i = \mathrm{lcm}(\sigma_{i_1}, \sigma_{i_2})$, which implies that for every $\sigma_{i_1} \in \sum_1$ and $\sigma_{i_2} \in \sum_2$, there are $\gcd(\sigma_{i_1}, \sigma_{i_2})$ occurrences of $\mathrm{lcm}(\sigma_{i_1}, \sigma_{i_2})$ in $\sum$. In the special case when $j_2 = 0$, every $\sigma_{i_1} \in \sum_1$ is present in $\sum$ and when $j_1 = 0$, every $\sigma_{i_2} \in \sum_2$ is present in $\sum$.

4.6   Appendix B

```
        PROGRAM MULT
C       PROGRAM TO COMPUTE A LOWER BOUND ON MULTIPLICATIVE
C       COMPLEXITY
C       FOR LENGTH N CYCLIC CONVOLUTION OVER THE GALOIS FIELD
C       GF(p)
        COMMON NN,P, CO3, XLOGP
        OPEN(1,FILE='#6:')
        READ(*,100) N,P
        XN=N
        XLOGP=ALOG(P)
        XO=1.0
        CO1=(2.0*P-1.0)/(P-1.0)
        CO2=P/(P-1.0)
        CO3=P/(2.0*P-1.0)
        XTOL=1.0E-7
        FTOL=1.0E-7
        NTOL=1000
        CALL NEWTON(XO,XTOL,FTOL,NTOL,IFLAG)
        IF(IFLAG.GT.1) GO TO 1
        A=((XN-XO+1.0)*CO2)/(P**XO)
        B=XO*CO1
        COMP=XN*CO1-(A+B-CO2)
        WRITE(1,200) N,XO,A,B,CO2,COMP
  100   FORMAT(I5,F4.2)
  200   FORMAT(I5,5F12.4)
    1   CONTINUE
        END
        SUBROUTINE NEWTON(XO,XTOL,FTOL,NTOL,IFLAG)
        COMMON XN,P,CO3,XLOGP
        IFLAG=0
        DO 2 I=1,NTOL
        FXO=P**XO-CO3*((XN-XO+1.0)*XLOGP+1.0)
        IF(ABS(FXO).LT.FTOL) GO TO 4
        DERIV=((P**XO)+CO3)*XLOGP
        IF(DERIV.EQ.0.0) GO TO 3
        DELTAX=FXO/DERIV
```

```
      XO=XO-DELTAX
      IF(ABS(DELTAX).LT.XTOL) GO TO 4
2     CONTINUE
3     IFLAG=2
4     CONTINUE
      RETURN
      END
```

Chapter 5

A New Class of Linear Codes

The subject of this chapter is the study of the algebraic structure of the class of linear codes obtained from bilinear cyclic and aperiodic convolution algorithms over the finite field of interest. As the binary codes are most extensively used in systems employing error-correcting codes, the emphasis in this chapter is on the study of binary codes and their properties. Such a class of linear codes possesses a number of interesting properties that can be used in digital communication systems. For a code of length n, obtained as a result of convolution (cyclic or aperiodic) of length $N = k + d - 1$, the decoder design *does not change significantly* when: (1) k and d are varied such that N and, consequently, n is a constant and (2) k is kept constant and d is varied, changing N and n accordingly. The encoding and decoding procedures for the linear codes are developed in this chapter.

The relationship between systems of bilinear forms and algebraic coding theory was established in Chapter 2. Via a particular system of bilinear forms (2.15) and its A-dual (2.22), it was shown that if $A^T(C^T \mathbf{z} \bullet B\mathbf{y})$ is an NC algorithm for computing the aperiodic convolution $\boldsymbol{\phi} = \mathbf{z} \odot \mathbf{y}$ of the length k sequence $\mathbf{z}$ and the length d sequence $\mathbf{y}$, then the generator matrix of the linear $(n, k, \bar{d})$ code is given by C and $\bar{d} \geq d$. Similarly, if $C(A\mathbf{x} \bullet B\mathbf{y})$ is an NC algorithm for computing the cyclic convolution of two length N ($N = k+d-1$) sequences $\mathbf{x}$ and $\mathbf{y}$, then the generator matrix of the linear $(n, k, \bar{d})$ code is obtained by simply deleting the last $(N - k)$ consecutive rows of C or the last $N - k$

consecutive columns of A. Without any loss of generality, we will focus our attention on the analysis of the linear code whose generator matrix is obtained by deleting the last $N - k$ consecutive columns of A.

In Chapter 3, efficient algorithms for cyclic and aperiodic convolution of sequences over finite fields were developed. The emphasis throughout is on minimizing the multiplicative complexity of the bilinear algorithms. Knowledge concerning the field of constants is incorporated in the design of various algorithms in order to reduce their multiplicative complexity. These algorithms could broadly be classified as: (1) multidimensional techniques and (2) CRT-based techniques. In the multidimensional technique, a one-dimensional convolution is converted into a two- or higher-dimensional convolution using an appropriate mapping of the indices. In the CRT-based technique, a long length convolution is broken up into a number of smaller length convolutions. The smaller length convolutions are computed using specialized algorithms (some of which are given in Appendix B of Chapter 3) and then the long length convolution is reconstructed using the CRT for polynomials. In this chapter, we are interested in studying the mathematical properties of the corresponding class of linear codes.

5.1   Multidimensional Algorithms and Codes

Let us consider the two multidimensional aperiodic convolution algorithms described in Chapter 3. In the first approach, the length k and length d sequences are extended to length N. For composite N, that is, $N = N_1 N_2$, one-dimensional sequences are converted into two-dimensional arrays of size $(N_1 \times N_2)$. These two-dimensional arrays are then convolved to obtain a $[(2N_1 - 1) \times (2N_2 - 1)]$-dimensional array. Finally, the aperiodic convolution coefficients are reconstructed using (3.19) with additions only. Alternately, a one-dimensional polynomial product can be converted into a multidimensional polynomial product for composite k and d having a common factor, that is, $k = k_1 n_1$, $d = d_1 n_1$, where $k - 1$ and $d - 1$ are the degrees of the polynomials $Z(u)$ and $Y(u)$, respectively. The first dimension corresponds to the product of two polynomials of degree $(n_1 - 1)$ each, while the second dimension correponds to a product of two polynomials of degrees $(k_1 - 1)$ and $(d_1 - 1)$. In either of the two approaches, it may be possible to further decompose the two-dimensional convolution into multidimensional polynomial products.

It is clear from the description here that the code corresponding to the multidimensional approach may be interpreted as a *product code* obtained as the product of the codes corresponding to each of the dimensions into which the original one-dimensional convolution is converted. The properties and algebraic structure of product codes are well known [5.1-5.3] and, therefore, the multidimensional convolution approach will not be pursued further in regard to linear code generation. We conclude our discussion, however, with an example illustrating the multidimensional approach for computing the polynomial product $\Phi(u) = Z(u)Y(u)$, where $k = d = 4$, and constructing the corresponding linear code over $GF(2)$.

Using the notation of Section 3.2.2, the one-dimensional polynomials are converted into two-dimensional polynomials as follows:

$$
\begin{aligned}
Z(u) = Z(u,v) &= (z_0 + z_2 v) + (z_1 + z_3 v)\, u \\
&= Z_0(v) + Z_1(v)\, u, \\
Y(u) = Y(u,v) &= (y_0 + y_2 v) + (y_1 + y_3 v) u \\
&= Y_0(v) + Y_1(v) u,
\end{aligned}
$$

where $v = u^2$. The two-dimensional polynomial product $\Phi(u,v) = Z(u,v)Y(u,v)$ is computed by recursively using the algorithm for multiplying two degree 1 polynomials given in Appendix B of Chapter 3. First we define

$$
\begin{aligned}
\mathbf{m_0} &= Z_0(v)Y_0(v), \\
\mathbf{m_1} &= Z_1(v)Y_1(v), \\
\mathbf{m_2} &= [Z_0(v) + Z_1(v)]\,[Y_0(v) + Y_1(v)];
\end{aligned}
$$

then, the two-dimensional polynomial $\Phi(u,v)$ is computed as

$$
\Phi(u,v) = \mathbf{m_0} + (\mathbf{m_0} + \mathbf{m_1} + \mathbf{m_2})\, u + \mathbf{m_1} u^2.
$$

The quantities $\mathbf{m_0}$, $\mathbf{m_1}$, and $\mathbf{m_2}$ are computed in the following manner. Let

$$
\begin{aligned}
m_0 &= z_0 y_0, \\
m_1 &= z_2 y_2, \\
m_2 &= (z_0 + z_2)\,(y_0 + y_2);
\end{aligned}
$$

then,

$$\mathbf{m_0} = m_0 + (m_0 + m_1 + m_2)\, v + m_1\, v^2.$$

Let

$$m_3 = z_1 y_1,$$
$$m_4 = z_3 y_3,$$
$$m_5 = (z_1 + z_3)\,(y_1 + y_3);$$

then,

$$\mathbf{m_1} = m_3 + (m_3 + m_4 + m_5)\, v + m_4\, v^2.$$

Let

$$m_6 = (z_0 + z_1)\,(y_0 + y_1),$$
$$m_7 = (z_2 + z_3)\,(y_2 + y_3),$$
$$m_8 = (z_0 + z_1 + z_2 + z_3)\,(y_0 + y_1 + y_2 + y_3);$$

then,

$$\mathbf{m_2} = m_6 + (m_6 + m_7 + m_8)\, v + m_7\, v^2.$$

The polynomial $\Phi(u)$ is obtained from $\Phi(u,v)$ by replacing v by u^2. The corresponding $(9,4,4)$ binary code is a two-dimensional product code, for which each of the constituent codes are $(3,2,2)$ codes. The entire encoding procedure can be considered as an arrangement of the four information bits into a square (2×2) matrix with a parity-check bit adjoined to each row and column of the matrix. For example, if $z_0 = 1$, $z_1 = 0$, $z_2 = 1$, and $z_3 = 0$, then the encoding process can be shown pictorially as

$$
\begin{array}{cc|c}
1 & 1 & 0 \\
0 & 0 & 0 \\
\hline
1 & 1 & 0
\end{array}
$$

It is interesting to observe that the above algorithm is used in the CRT-based convolution approach for the computation $\Phi_i(u) \equiv Z_i(u)\, Y_i(u) \bmod P_i(u)$, if $\deg [P_i(u)] = 4$ and $P_i(u)$ is irreducible.

5.2 CRT Algorithms and Codes

The CRT-based aperiodic convolution algorithms (with wraparound) were derived in Section 3.2.3. In this section, we study these algorithms further with the objective of establishing the related class of linear codes and their properties.

The CRT-based algorithm for computing the product of two polynomials $Z(u)$ and $Y(u)$ of degrees $k-1$ and $d-1$, respectively, consists of selecting a polynomial $P(u)$ of degree D and a scalar s such that $N = D+s$, $N = k+d-1$. The polynomial $\Phi(u) = Z(u)Y(u)$ is then obtained (using only additions) from the polynomial $\Phi'(u)$ computed as $\Phi'(u) \equiv \Phi(u) \bmod P(u)$ and the s wraparound coefficients computed as $\hat{\Phi}(u) \equiv \hat{Z}(u)\hat{Y}(u) \bmod u^s$. The polynomials $\hat{Z}(u)$ and $\hat{Y}(u)$ are the reciprocal polynomials corresponding to $Z(u)$ and $Y(u)$, respectively. The polynomial $P(u)$ is chosen as the product of L relatively prime polynomials $P_i(u)$, $i = 1, 2, \ldots, L$, and computation of $\Phi'(u)$ is performed in three steps: (1) reduce the polynomials $Z(u)$ and $Y(u)$ modulo $P_i(u)$, to obtain $Z_i(u) \equiv Z(u) \bmod P_i(u)$, $Y_i(u) \equiv Y(u) \bmod P_i(u)$, $i = 1, 2, \ldots, L$; (2) compute $\Phi_i(u) \equiv Z_i(u)Y_i(u) \bmod P_i(u)$, $i = 1, 2, \ldots, L$; and (3) use the CRT to reconstruct $\Phi'(u)$ from the $\Phi_i(u)$, $i = 1, 2, \ldots, L$.

If $P(u)$ is selected to be $(u^D - 1)$, then $\phi_0, \phi_1, \ldots, \phi_{s/2}$ and $\phi_{D+s/2}$, $\phi_{D+1+s/2}, \ldots, \phi_{N-1}$, for s even, or $\phi_0, \phi_1, \ldots, \phi_{(s-1)/2}$ and $\phi_{D+(s-1)/2}$, $\phi_{D+1+(s-1)/2}, \ldots, \phi_{N-1}$, for s odd, can be computed as the wraparound coefficients in place of ϕ_D, $\phi_{D+1}, \ldots, \phi_{N-1}$, as stated above (cf. Lemma 3.3). Also, if $P(u)$ is selected to be $(u^N - 1)$, then the CRT-based aperiodic convolution algorithm can be derived from the CRT-based cyclic convolution algorithm described in Section 3.1.

In the following section, several examples of algorithms for aperiodic convolution are presented, along with the corresponding linear codes generated. The various properties of such code families are also described.

5.2.1 CRT-Based Convolution Algorithms over GF(2) and the Related Codes

Appendix C of Chapter 3 lists a number of monic polynomials over $GF(2)$ up to degree 5. For each degree, only those polynomials are listed which are not products of two distinct polynomials of lower degree. For example, $(u^3 + u^2)$ is not listed as a degree 3 polynomial;

this is to help in the choice of the polynomial $P(u)$ as a product of relatively prime polynomials $P_i(u)$, $i = 1, 2, \ldots, L$. For a given degree of the polynomial $P(u)$, its factors are selected in such a way that the multiplicative complexity of the computation $Z(u)Y(u) \bmod P(u)$ is as low as possible.

Given below is a list of polynomials of degree D up to 15 obtained by selecting appropriate polynomials from the list provided:

$D = 3$: $P(u) = (u + 1)\,(u^2 + u + 1)$

$D = 4$: $P(u) = u(u + 1)\,(u^2 + u + 1)$

$D = 5$: $P(u) = u(u^2 + 1)\,(u^2 + u + 1)$

$D = 6$: $P(u) = u^2\,(u^2 + 1)\,(u^2 + u + 1)$

$D = 7$: $P(u) = u(u + 1)\,(u^2 + u + 1)\,(u^3 + u^2 + 1)$

$D = 8$: $P(u) = u(u^2 + 1)\,(u^2 + u + 1)\,(u^3 + u^2 + 1)$

$D = 9$: $P(u) = u^2(u^2 + 1)\,(u^2 + u + 1)\,(u^3 + u^2 + 1)$

$D = 10$: $P(u) = u(u + 1)\,(u^2 + u + 1)\,(u^3 + u + 1)\,(u^3 + u^2 + 1)$

$D = 11$: $P(u) = u(u^2 + 1)\,(u^2 + u + 1)\,(u^3 + u + 1)\,(u^3 + u^2 + 1)$

$D = 12$: $P(u) = u^2(u^2 + 1)\,(u^2 + u + 1)(u^3 + u + 1)$
$\qquad\qquad\qquad \cdot (u^3 + u^2 + 1)$

$D = 13$: $P(u) = u^3(u^2 + 1)\,(u^2 + u + 1)\,(u^3 + u + 1)$
$\qquad\qquad\qquad \cdot (u^3 + u^2 + 1)$

$D = 14$: $P(u) = u^3(u^3 + u^2 + u + 1)\,(u^2 + u + 1)\,(u^3 + u + 1)$
$\qquad\qquad\qquad \cdot (u^3 + u^2 + 1)$

$D = 15$: $P(u) = u(u^2 + 1)\,(u^2 + u + 1)\,(u^3 + u + 1)\,(u^3 + u^2 + 1)$
$\qquad\qquad\qquad \cdot (u^4 + u + 1)$

Note that the choice of the polynomial $P(u)$ is not unique for a given degree D. For example, when $D = 5$, $P(u)$ can have either one of the two forms, $P(u) = u^2(u+1)\,(u^2+u+1)$ or $P(u) = u(u^2+1)\,(u^2+u+1)$. The multiplicative complexity of the procedure $Z(u)Y(u) \bmod P(u)$ is, however, the *same* for all such choices. In the following example, we derive bilinear convolution algorithms of length 6 and 16 and the corresponding binary codes.

EXAMPLE 1.   Consider the polynomial product $Z(u)Y(u)$, where $Z(u) = \sum_{i=0}^{3} z_i u^i$ and $Y(u) = \sum_{i=0}^{2} y_i u^i$. Clearly, $k = 4$ and $d = 3$. We choose $P(u) = u(u^2 + 1)\,(u^2 + u + 1)$ and $s = 1$ in order to compute the aperiodic convolution $\Phi(u) = Z(u)Y(u)$. Let $P_1(u) = u$, $P_2(u) = (u^2 + 1)$, and $P_3(u) = (u^2 + u + 1)$.

Reducing the polynomials $Z(u)$ and $Y(u)$ modulo each of the $P_i(u)$, we obtain

$$
\begin{aligned}
Z_1(u) &\equiv Z(u) \bmod u \\
&= z_0, \\
Y_1(u) &\equiv Y(u) \bmod u \\
&= y_0.
\end{aligned}
$$

Let $m_0 = z_0 y_0$. Then, $\Phi_1(u) \equiv \Phi(u) \bmod u$ is equal to m_0. Similarly,

$$
\begin{aligned}
Z_2(u) &\equiv Z(u) \bmod (u^2 + 1) \\
&= (z_0 + z_2) + (z_1 + z_3)\, u, \\
Y_2(u) &\equiv Y(u) \bmod (u^2 + 1) \\
&= (y_0 + y_2) + y_1\, u.
\end{aligned}
$$

Let

$$
\begin{aligned}
m_1 &= (z_0 + z_2)(y_0 + y_2), \\
m_2 &= (z_0 + z_1 + z_2 + z_3)(y_0 + y_1 + y_2), \\
m_3 &= (z_1 + z_3)\, y_1;
\end{aligned}
$$

then,

$$
\begin{aligned}
\Phi_2(u) &\equiv \Phi(u) \bmod (u^2 + 1) \\
&= (m_1 + m_3) + (m_1 + m_2 + m_3)\, u.
\end{aligned}
$$

Also,

$$
\begin{aligned}
Z_3(u) &\equiv Z(u) \bmod (u^2 + u + 1) \\
&= (z_0 + z_2 + z_3) + (z_1 + z_2)\, u, \\
Y_3(u) &\equiv Y(u) \bmod (u^2 + u + 1) \\
&= (y_0 + y_2) + (y_1 + y_2)\, u.
\end{aligned}
$$

Let

$$
\begin{aligned}
m_4 &= (z_0 + z_2 + z_3)(y_0 + y_2), \\
m_5 &= (z_0 + z_1 + z_3)(y_0 + y_1), \\
m_6 &= (z_1 + z_2)(y_1 + y_2);
\end{aligned}
$$

then,

$$\begin{aligned}
\Phi_3(u) &\equiv \Phi(u) \bmod (u^2 + u + 1)\\
&= (m_4 + m_6) + (m_4 + m_5)\, u.
\end{aligned}$$

The polynomial $\Phi'(u) \equiv \Phi(u) \bmod P(u)$ can be recovered from the polynomials $\Phi_i(u), i = 1, 2, 3$, using the CRT, i.e.,

$$\Phi(u) \equiv \left[\sum_{i=1}^{3} S_i(u)\, \Phi_i(u)\right] \bmod P(u).$$

The polynomials $S_i(u)$, $i = 1, 2, 3$, are found using Euclid's algorithm and are $S_1(u) = (u^4 + u^3 + u + 1)$, $S_2(u) = (u^3 + u^2 + u)$, and $S_3(u) = (u^4 + u^2)$. If $\Phi'(u) \equiv (\phi'_0 + \phi'_1\, u + \phi'_2\, u^2 + \phi'_3\, u^3 + \phi'_4\, u^4) \bmod P(u)$, then it can be shown that

$$\begin{aligned}
\phi'_0 &= m_0,\\
\phi'_1 &= m_0 + m_1 + m_3 + m_4 + m_5,\\
\phi'_2 &= m_2 + m_5 + m_6,\\
\phi'_3 &= m_0 + m_2 + m_4 + m_5,\\
\phi'_4 &= m_0 + m_1 + m_2 + m_3 + m_5 + m_6.
\end{aligned} \qquad (5.1)$$

Let $m_7 = z_3 y_2$, the polynomial product $\Phi(u) = Z(u)Y(u) = \phi_0 + \phi_1\, u + \phi_2\, u^2 + \phi_3\, u^3 + \phi_4\, u^4 + \phi_5\, u^5$ can be computed from (5.1), and the coefficients ϕ_i, $i = 0, 1, \ldots, 5$, are given by

$$\begin{aligned}
\phi_0 &= m_0,\\
\phi_1 &= m_0 + m_1 + m_3 + m_4 + m_5 + m_7,\\
\phi_2 &= m_2 + m_5 + m_6 + m_7,\\
\phi_3 &= m_0 + m_2 + m_4 + m_5,\\
\phi_4 &= m_0 + m_1 + m_2 + m_3 + m_5 + m_6 + m_7,\\
\phi_5 &= m_7.
\end{aligned}$$

The bilinear algorithm for the computation of the above aperiodic convolution is given by

$$
\begin{bmatrix} \phi_0 \\ \phi_1 \\ \phi_2 \\ \phi_3 \\ \phi_4 \\ \phi_5 \end{bmatrix} = \begin{bmatrix} 1 & 0 & 0 & 0 & 0 & 0 & 0 & 0 \\ 1 & 1 & 0 & 1 & 1 & 1 & 0 & 1 \\ 0 & 0 & 1 & 0 & 0 & 1 & 1 & 1 \\ 1 & 0 & 1 & 0 & 1 & 1 & 0 & 0 \\ 1 & 1 & 1 & 1 & 0 & 1 & 1 & 1 \\ 0 & 0 & 0 & 0 & 0 & 0 & 0 & 1 \end{bmatrix} \left(\begin{bmatrix} 1 & 0 & 0 & 0 \\ 1 & 0 & 1 & 0 \\ 1 & 1 & 1 & 1 \\ 0 & 1 & 0 & 1 \\ 1 & 0 & 1 & 1 \\ 1 & 1 & 0 & 1 \\ 0 & 1 & 1 & 0 \\ 0 & 0 & 0 & 1 \end{bmatrix} \begin{bmatrix} z_0 \\ z_1 \\ z_2 \\ z_3 \end{bmatrix} \right.
$$

$$
\left. \bullet \begin{bmatrix} 1 & 0 & 0 \\ 1 & 0 & 1 \\ 1 & 1 & 1 \\ 0 & 1 & 0 \\ 1 & 0 & 1 \\ 1 & 1 & 0 \\ 0 & 1 & 1 \\ 0 & 0 & 1 \end{bmatrix} \begin{bmatrix} y_0 \\ y_1 \\ y_2 \end{bmatrix} \right)
$$

$$
= A^T(C^T \mathbf{z} \bullet B\mathbf{y}).
$$

The generator matrix for the corresponding (8,4,3) single error- correcting code is

$$
C = \begin{bmatrix} 1 & 1 & 1 & 0 & 1 & 1 & 0 & 0 \\ 0 & 0 & 1 & 1 & 0 & 1 & 1 & 0 \\ 0 & 1 & 1 & 0 & 1 & 0 & 1 & 0 \\ 0 & 0 & 1 & 1 & 1 & 1 & 0 & 1 \end{bmatrix}.
$$

Note that it is not possible to find a code with the same value of n and d for which k is larger.

EXAMPLE 2. Consider the polynomial product $Z(u)Y(u)$, where $Z(u) = \sum_{i=0}^{7} z_i u^i$ and $Y(u) = \sum_{i=0}^{8} y_i u^i$. Clearly, $k = 8$ and $d = 9$. Choosing $P(u) = u^3(u^2 + 1)(u^3 + u + 1)(u^3 + u^2 + 1)(u^2 + u + 1)$ and $s = 3$, it can be shown that the polynomial product $\Phi(u) = Z(u)Y(u)$ can be computed using the algorithm $A^T(C^T \mathbf{z} \bullet B\mathbf{y})$, where the matrices A, B, and C are given by

$$A = \begin{bmatrix}
1 & 1 & 1 & 1 & 1 & 1 & 0 & 1 & 1 & 1 & 1 & 1 & 1 & 0 & 0 & 0 \\
0 & 1 & 1 & 0 & 1 & 1 & 0 & 0 & 1 & 1 & 0 & 1 & 1 & 0 & 0 & 0 \\
0 & 0 & 1 & 0 & 0 & 1 & 0 & 0 & 0 & 1 & 0 & 0 & 0 & 0 & 0 & 0 \\
0 & 1 & 0 & 0 & 1 & 0 & 0 & 0 & 1 & 0 & 0 & 1 & 1 & 0 & 0 & 0 \\
0 & 0 & 1 & 0 & 0 & 1 & 0 & 0 & 0 & 1 & 0 & 0 & 0 & 0 & 0 & 0 \\
0 & 0 & 0 & 1 & 1 & 1 & 0 & 0 & 0 & 0 & 1 & 1 & 1 & 0 & 0 & 0 \\
0 & 0 & 0 & 1 & 1 & 1 & 0 & 0 & 0 & 0 & 1 & 1 & 1 & 0 & 0 & 0 \\
0 & 0 & 0 & 1 & 0 & 1 & 1 & 1 & 1 & 1 & 0 & 1 & 1 & 0 & 0 & 0 \\
0 & 0 & 0 & 0 & 1 & 1 & 0 & 0 & 0 & 0 & 0 & 1 & 1 & 0 & 0 & 0 \\
0 & 0 & 0 & 1 & 0 & 1 & 0 & 0 & 0 & 0 & 1 & 1 & 0 & 0 & 0 & 0 \\
0 & 0 & 0 & 1 & 1 & 0 & 0 & 0 & 0 & 0 & 1 & 0 & 1 & 0 & 0 & 0 \\
0 & 0 & 0 & 0 & 1 & 1 & 1 & 1 & 0 & 1 & 0 & 1 & 1 & 0 & 0 & 0 \\
0 & 0 & 0 & 0 & 0 & 1 & 1 & 1 & 1 & 0 & 1 & 0 & 0 & 0 & 0 & 0 \\
0 & 0 & 0 & 1 & 0 & 0 & 0 & 1 & 1 & 1 & 1 & 1 & 1 & 0 & 0 & 0 \\
0 & 0 & 0 & 0 & 1 & 0 & 0 & 0 & 1 & 1 & 1 & 1 & 1 & 0 & 0 & 0 \\
0 & 0 & 0 & 1 & 1 & 0 & 0 & 1 & 0 & 0 & 0 & 0 & 0 & 0 & 0 & 0 \\
0 & 0 & 0 & 1 & 0 & 1 & 1 & 0 & 0 & 1 & 0 & 1 & 1 & 0 & 0 & 0 \\
0 & 0 & 0 & 0 & 0 & 0 & 0 & 1 & 0 & 1 & 1 & 0 & 1 & 0 & 0 & 0 \\
0 & 0 & 0 & 1 & 0 & 0 & 0 & 0 & 1 & 0 & 0 & 0 & 1 & 0 & 0 & 0 \\
0 & 0 & 0 & 1 & 1 & 0 & 1 & 0 & 0 & 1 & 1 & 0 & 0 & 0 & 0 & 0 \\
0 & 0 & 0 & 0 & 1 & 1 & 1 & 1 & 1 & 0 & 0 & 0 & 0 & 0 & 0 & 0 \\
0 & 0 & 0 & 0 & 1 & 0 & 1 & 0 & 1 & 1 & 1 & 0 & 1 & 0 & 0 & 0 \\
0 & 0 & 0 & 1 & 0 & 1 & 0 & 1 & 1 & 1 & 1 & 0 & 0 & 0 & 0 & 0 \\
0 & 0 & 0 & 1 & 1 & 1 & 1 & 1 & 1 & 0 & 1 & 1 & 1 & 1 & 1 & 1 \\
0 & 0 & 0 & 0 & 1 & 0 & 0 & 1 & 0 & 0 & 0 & 1 & 0 & 0 & 1 & 0 \\
0 & 0 & 0 & 0 & 1 & 0 & 0 & 1 & 0 & 0 & 0 & 1 & 0 & 0 & 1 & 0 \\
0 & 0 & 0 & 1 & 0 & 0 & 1 & 0 & 0 & 0 & 1 & 0 & 0 & 1 & 0 & 0 \\
0 & 0 & 0 & 1 & 0 & 0 & 1 & 0 & 0 & 0 & 1 & 0 & 0 & 1 & 0 & 0
\end{bmatrix}.$$

$$B = \begin{bmatrix}
1 & 0 & 0 & 0 & 0 & 0 & 0 & 0 & 0 \\
0 & 1 & 0 & 0 & 0 & 0 & 0 & 0 & 0 \\
0 & 0 & 1 & 0 & 0 & 0 & 0 & 0 & 0 \\
1 & 1 & 0 & 0 & 0 & 0 & 0 & 0 & 0 \\
1 & 0 & 1 & 0 & 0 & 0 & 0 & 0 & 0 \\
1 & 0 & 1 & 0 & 1 & 0 & 1 & 0 & 1 \\
0 & 1 & 0 & 1 & 0 & 1 & 0 & 1 & 0 \\
1 & 1 & 1 & 1 & 1 & 1 & 1 & 1 & 1 \\
1 & 0 & 1 & 1 & 0 & 1 & 1 & 0 & 1 \\
0 & 1 & 1 & 0 & 1 & 1 & 0 & 1 & 1 \\
1 & 1 & 0 & 1 & 1 & 0 & 1 & 1 & 0 \\
1 & 0 & 0 & 1 & 0 & 1 & 1 & 1 & 0 \\
0 & 1 & 0 & 1 & 1 & 1 & 0 & 0 & 1 \\
0 & 0 & 1 & 0 & 1 & 1 & 1 & 0 & 0 \\
1 & 1 & 0 & 0 & 1 & 0 & 1 & 1 & 1 \\
0 & 1 & 1 & 1 & 0 & 0 & 1 & 0 & 1 \\
1 & 0 & 1 & 1 & 1 & 0 & 0 & 1 & 0 \\
1 & 0 & 0 & 1 & 1 & 1 & 0 & 1 & 0 \\
0 & 1 & 0 & 0 & 1 & 1 & 1 & 0 & 1 \\
0 & 0 & 1 & 1 & 1 & 0 & 1 & 0 & 0 \\
1 & 1 & 0 & 1 & 0 & 0 & 1 & 1 & 1 \\
0 & 1 & 1 & 1 & 0 & 1 & 0 & 0 & 1 \\
1 & 0 & 1 & 0 & 0 & 1 & 1 & 1 & 0 \\
0 & 0 & 0 & 0 & 0 & 0 & 0 & 0 & 1 \\
0 & 0 & 0 & 0 & 0 & 0 & 0 & 1 & 1 \\
0 & 0 & 0 & 0 & 0 & 0 & 0 & 1 & 0 \\
0 & 0 & 0 & 0 & 0 & 0 & 1 & 0 & 1 \\
0 & 0 & 0 & 0 & 0 & 0 & 1 & 0 & 0
\end{bmatrix},$$

and

$$
C =
\begin{bmatrix}
1 & 0 & 0 & 0 & 0 & 0 & 0 & 0 \\
0 & 1 & 0 & 0 & 0 & 0 & 0 & 0 \\
0 & 0 & 1 & 0 & 0 & 0 & 0 & 0 \\
1 & 1 & 0 & 0 & 0 & 0 & 0 & 0 \\
1 & 0 & 1 & 0 & 0 & 0 & 0 & 0 \\
1 & 0 & 1 & 0 & 1 & 0 & 1 & 0 \\
0 & 1 & 0 & 1 & 0 & 1 & 0 & 1 \\
1 & 1 & 1 & 1 & 1 & 1 & 1 & 1 \\
1 & 0 & 1 & 1 & 0 & 1 & 1 & 0 \\
0 & 1 & 1 & 0 & 1 & 1 & 0 & 1 \\
1 & 1 & 0 & 1 & 1 & 0 & 1 & 1 \\
1 & 0 & 0 & 1 & 0 & 1 & 1 & 1 \\
0 & 1 & 0 & 1 & 1 & 1 & 0 & 0 \\
0 & 0 & 1 & 0 & 1 & 1 & 1 & 0 \\
1 & 1 & 0 & 0 & 1 & 0 & 1 & 1 \\
0 & 1 & 1 & 1 & 0 & 0 & 1 & 0 \\
1 & 0 & 1 & 1 & 1 & 0 & 0 & 1 \\
1 & 0 & 0 & 1 & 1 & 1 & 0 & 1 \\
0 & 1 & 0 & 0 & 1 & 1 & 1 & 0 \\
0 & 0 & 1 & 1 & 1 & 0 & 1 & 0 \\
1 & 1 & 0 & 1 & 0 & 0 & 1 & 1 \\
0 & 1 & 1 & 1 & 0 & 1 & 0 & 0 \\
1 & 0 & 1 & 0 & 0 & 1 & 1 & 1 \\
0 & 0 & 0 & 0 & 0 & 0 & 0 & 1 \\
0 & 0 & 0 & 0 & 0 & 0 & 1 & 1 \\
0 & 0 & 0 & 0 & 0 & 0 & 1 & 0 \\
0 & 0 & 0 & 0 & 0 & 1 & 0 & 1 \\
0 & 0 & 0 & 0 & 0 & 1 & 0 & 0 \\
\end{bmatrix}^{T} .
$$

The matrix C is also the generator matrix of a (28, 8, 9) four error-correcting code. The above matrices can be easily modified to accomodate the case when $k = 6$ and $d = 11$. The generator matrix of the corresponding (28, 6, 11) five error-correcting code can be shown to be

$$C = \begin{bmatrix} 1 & 0 & 0 & 0 & 0 & 0 \\ 0 & 1 & 0 & 0 & 0 & 0 \\ 0 & 0 & 1 & 0 & 0 & 0 \\ 1 & 1 & 0 & 0 & 0 & 0 \\ 1 & 0 & 1 & 0 & 0 & 0 \\ 1 & 0 & 1 & 0 & 1 & 0 \\ 0 & 1 & 0 & 1 & 0 & 1 \\ 1 & 1 & 1 & 1 & 1 & 1 \\ 1 & 0 & 1 & 1 & 0 & 1 \\ 0 & 1 & 1 & 0 & 1 & 1 \\ 1 & 1 & 0 & 1 & 1 & 0 \\ 1 & 0 & 0 & 1 & 0 & 1 \\ 0 & 1 & 0 & 1 & 1 & 1 \\ 0 & 0 & 1 & 0 & 0 & 1 \\ 1 & 1 & 0 & 0 & 1 & 0 \\ 0 & 1 & 1 & 1 & 0 & 0 \\ 1 & 0 & 1 & 1 & 1 & 0 \\ 1 & 0 & 0 & 1 & 1 & 1 \\ 0 & 1 & 0 & 0 & 1 & 1 \\ 0 & 0 & 1 & 1 & 1 & 0 \\ 1 & 1 & 0 & 1 & 0 & 0 \\ 0 & 1 & 1 & 1 & 0 & 1 \\ 1 & 0 & 1 & 0 & 0 & 1 \\ 0 & 0 & 0 & 0 & 0 & 1 \\ 0 & 0 & 0 & 0 & 1 & 1 \\ 0 & 0 & 0 & 0 & 1 & 0 \\ 0 & 0 & 0 & 1 & 0 & 1 \\ 0 & 0 & 0 & 1 & 0 & 0 \end{bmatrix}^{T}.$$

Similarly, the generator matrix of the (28,10,7) three error-correcting code can be obtained as

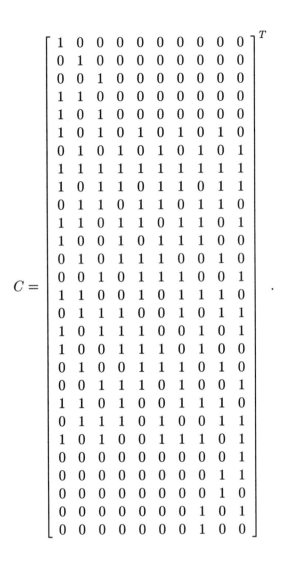

$$
C = \begin{bmatrix}
1 & 0 & 0 & 0 & 0 & 0 & 0 & 0 & 0 & 0 \\
0 & 1 & 0 & 0 & 0 & 0 & 0 & 0 & 0 & 0 \\
0 & 0 & 1 & 0 & 0 & 0 & 0 & 0 & 0 & 0 \\
1 & 1 & 0 & 0 & 0 & 0 & 0 & 0 & 0 & 0 \\
1 & 0 & 1 & 0 & 0 & 0 & 0 & 0 & 0 & 0 \\
1 & 0 & 1 & 0 & 1 & 0 & 1 & 0 & 1 & 0 \\
0 & 1 & 0 & 1 & 0 & 1 & 0 & 1 & 0 & 1 \\
1 & 1 & 1 & 1 & 1 & 1 & 1 & 1 & 1 & 1 \\
1 & 0 & 1 & 1 & 0 & 1 & 1 & 0 & 1 & 1 \\
0 & 1 & 1 & 0 & 1 & 1 & 0 & 1 & 1 & 0 \\
1 & 1 & 0 & 1 & 1 & 0 & 1 & 1 & 0 & 1 \\
1 & 0 & 0 & 1 & 0 & 1 & 1 & 1 & 0 & 0 \\
0 & 1 & 0 & 1 & 1 & 1 & 0 & 0 & 1 & 0 \\
0 & 0 & 1 & 0 & 1 & 1 & 1 & 0 & 0 & 1 \\
1 & 1 & 0 & 0 & 1 & 0 & 1 & 1 & 1 & 0 \\
0 & 1 & 1 & 1 & 0 & 0 & 1 & 0 & 1 & 1 \\
1 & 0 & 1 & 1 & 1 & 0 & 0 & 1 & 0 & 1 \\
1 & 0 & 0 & 1 & 1 & 1 & 0 & 1 & 0 & 0 \\
0 & 1 & 0 & 0 & 1 & 1 & 1 & 0 & 1 & 0 \\
0 & 0 & 1 & 1 & 1 & 0 & 1 & 0 & 0 & 1 \\
1 & 1 & 0 & 1 & 0 & 0 & 1 & 1 & 1 & 0 \\
0 & 1 & 1 & 1 & 0 & 1 & 0 & 0 & 1 & 1 \\
1 & 0 & 1 & 0 & 0 & 1 & 1 & 1 & 0 & 1 \\
0 & 0 & 0 & 0 & 0 & 0 & 0 & 0 & 0 & 1 \\
0 & 0 & 0 & 0 & 0 & 0 & 0 & 0 & 1 & 1 \\
0 & 0 & 0 & 0 & 0 & 0 & 0 & 0 & 1 & 0 \\
0 & 0 & 0 & 0 & 0 & 0 & 0 & 1 & 0 & 1 \\
0 & 0 & 0 & 0 & 0 & 0 & 0 & 1 & 0 & 0
\end{bmatrix}^{T}.
$$

Table 5.1 lists selected binary linear codes that can be obtained using the CRT-based convolution algorithm for lengths up to 30 along with their parameters. Note that all the codes marked with an asterisk are either the same as the *best known codes* or very close to the already known best codes [5.1, 5.4]. The list of binary codes that can be derived using the CRT-based aperiodic convolution algorithm is quite

Table 5.1: Selected Binary Linear Codes Obtained From the Aperiodic Convolution Algorithm.

Length of Conv. N	(n, k, d) Parameters
5	$(6,3, 3)^*$
6	$(8, 4, 3)^*$, $(8, 2, 5)$
7	$(10, 5, 3)$, $(10, 3, 5)^*$
8	$(12, 6, 3)$, $(12, 4, 5)^*$, $(12, 2, 7)^*$
9	$(14, 7, 3)$, $(14, 5, 5)^*$, $(14, 3, 7)^*$
10	$(16,8, 3)$, $(16, 6, 5)$, $(16, 4, 7)^*$
11	$(18, 9, 3)$, $(18, 7, 5)$, $(18, 5, 7)$, $(18, 3, 9)^*$
12	$(20, 10, 3)$, $(20, 8, 5)$, $(20, 6, 7)$, $(20, 4, 9)^*$
13	$(22, 9, 5)$, $(22, 7, 7)$, $(22, 5, 9)^*$, $(22, 3, 11)^*$
14	$(24, 10, 5)$, $(24, 8, 7)$, $(24, 6, 9)^*$, $(24, 4, 11)^*$
15	$(26, 11, 5)$, $(26, 9, 7)$, $(26, 7, 9)^*$, $(26, 5, 11)^*$, $(26, 3, 13)^*$
16	$(28, 10, 7)$, $(28, 8, 9)^*$, $(28, 6, 11)^*$, $(28, 4, 13)^*$
17	$(30, 11, 7)$, $(30, 9 ,9)$, $(30, 7, 11)^*$, $(30, 5, 13)^*$, $(30, 3, 15)$

large, and codes have been derived with length up to 100 and minimum distance up to 41.

There are certain advantages to this approach for linear code generation, some of which were mentioned briefly in Section 3.2.3. We elaborate on these here:

1. Knowing k and d, we can derive an error-correcting code for these parameters. A wide choice is possible because the codes are *not* restricted to be cyclic.

2. The codes have the unique property that $k+d-1 = $ constant; this provides a particular family of codes of length n. For example, the $(26, 9, 7)$, $(26, 7, 9)$, and $(26, 5, 11)$ codes belong to the same family.

There is, therefore, a systematic way to decrease (increase) k in order to increase (decrease) d for a given n.

3. The columns of the generator matrix C correspond to (a) multiplications required to compute the polynomial product $Z(u)Y(u)$ mod $P(u)$ or, (b) multiplications required to compute the polynomial product $\hat{Z}(u)\hat{Y}(u)$ mod u^s (wraparound). We can decrease k by suitably shortening the columns of the matrix C. The columns corresponding to (a) are shortened from the *bottom*, and the columns corresponding to (b) are shortened from the *top*. For example, the generator matrix of the (12, 6, 3) code is given by

$$
C = \begin{bmatrix}
1 & 0 & 1 & 1 & 0 & 1 & 1 & 0 & 1 & 0 & 0 & 0 \\
0 & 1 & 1 & 0 & 1 & 1 & 0 & 1 & 1 & 0 & 0 & 0 \\
0 & 0 & 0 & 1 & 0 & 1 & 1 & 1 & 0 & 0 & 0 & 0 \\
0 & 0 & 0 & 0 & 1 & 1 & 1 & 0 & 1 & 0 & 0 & 0 \\
0 & 0 & 0 & 1 & 0 & 1 & 0 & 1 & 1 & 0 & 1 & 1 \\
0 & 0 & 0 & 0 & 1 & 1 & 1 & 1 & 0 & 1 & 0 & 1
\end{bmatrix}. \tag{5.2}
$$

The last three columns of the matrix C correspond to the wraparound. The generator matrix of the (12, 4, 5) code can, therefore, be obtained by appropriately shortening the columns of the matrix C as shown in (5.2) by the dashed line. The generator matrix of the (12, 4, 5) code is

$$
C = \begin{bmatrix}
1 & 0 & 1 & 1 & 0 & 1 & 1 & 0 & 1 & 0 & 0 & 0 \\
0 & 1 & 1 & 0 & 1 & 1 & 0 & 1 & 1 & 0 & 0 & 0 \\
0 & 0 & 0 & 1 & 0 & 1 & 1 & 1 & 0 & 0 & 1 & 1 \\
0 & 0 & 0 & 0 & 1 & 1 & 1 & 0 & 1 & 1 & 0 & 1
\end{bmatrix}. \tag{5.3}
$$

Thus, a change in k *does not imply* a significant change in the encoding procedure. It is also important to mention here that a change in k *does not significantly alter* the decoding procedure. This point is illustrated further by the generalized decoding procedure to be presented in Section 5.4.

4. For a given value of k, it is possible to decrease (increase) the design minimum distance d of the code by simply deleting (adding) columns of (to) the generator matrix, thereby leading to an appropriate decrease (increase) in the length n of the code. The encoder/decoder design for all such codes remains *essentially unaltered*. The system designer can, therefore, incorporate a wide range of different codes in a *single* encoder/decoder structure. This property will be established in Section 5.4.

5. The above two points indicate that the error-correcting capability $\lfloor (d-1)/2 \rfloor$, of the codes generated can be altered quite easily, a feature which can be valuable in a *fluctuating noise environment*. Now, in many cases, it is necessary to change code families and, consequently, the entire encoding/decoding procedure to accomodate such situations.

5.2.2   CRT-Based Convolution Algorithms over GF(3) and the Related Codes

Appendix C of Chapter 3 lists a number of monic polynomials over $GF(3)$ up to degree 3. Once again, for each degree, only those polynomials are listed which are not products of two distinct polynomials of lower degree. Based on these polynomials, a list of polynomials of degree up to 10 is given below. The factors are selected so as to minimize the multiplicative complexity of computing the polynomial product $Z(u)Y(u) \bmod P(u)$:

$D = 3$: $P(u) = u(u+1)\,(u+2)$
$D = 4$: $P(u) = (u+1)\,(u+2)(u^2+u+2)$
$D = 5$: $P(u) = u(u+1)\,(u+2)\,(u^2+u+2)$
$D = 6$: $P(u) = u^2(u+1)\,(u+2)\,(u^2+u+2)$
$D = 7$: $P(u) = u(u+1)\,(u+2)\,(u^2+u+2)\,(u^2+2u+2)$
$D = 8$: $P(u) = u(u+1)\,(u^2+1)\,(u^2+u+2)\,(u^2+2u+1)$
$D = 9$: $P(u) = u(u+1)\,(u+2)\,(u^2+1)\,(u^2+u+2)$
$\qquad\qquad (u^2+2u+2)$
$D = 10$: $P(u) = u^2(u+1)\,(u+2)\,(u^2+1)\,(u^2+u+2)$
$\qquad\qquad (u^2+2u+2)$

In the following example, we derive a bilinear convolution algorithm of length 8 and the corresponding code over $GF(3)$.

EXAMPLE 3.   Consider the polynomial product $Z(u)Y(u)$, where $Z(u) = \sum_{i=0}^{3} z_i u^i$ and $Y(u) = \sum_{i=0}^{4} y_i u^i$. Clearly, $k = 4$ and $d = 5$. We choose $P(u) = u(u + 1) \; (u + 2) \; (u^2 + u + 2) \; (u^2 + 2u + 2)$ and $s = 1$ to compute the aperiodic convolution $\Phi(u) = Z(u)Y(u)$. Let $P_1(u) = u$, $P_2(u) = (u + 1)$, $P_3(u) = (u + 2)$, $P_4(u) = (u^2 + u + 2)$, and $P_5(u) = (u^2 + 2u + 2)$. Reducing the polynomials $Z(u)$ and $Y(u)$ modulo each of the $P_i(u)$, we obtain

$$
\begin{aligned}
Z_1(u) &\equiv Z(u) \bmod u \\
&= z_0, \\
Y_1(u) &\equiv Y(u) \bmod u \\
&= y_0.
\end{aligned}
$$

Let $m_0 = z_0 y_0$. Similarly,

$$
\begin{aligned}
Z_2(u) &\equiv Z(u) \bmod (u + 1) \\
&= z_0 + 2z_1 + z_2 + 2z_3, \\
Y_2(u) &\equiv Y(u) \bmod (u + 1) \\
&= y_0 + 2y_1 + y_2 + 2y_3 + y_4.
\end{aligned}
$$

Let
$$
m_1 = (z_0 + 2z_1 + z_2 + 2z_3) \, (y_0 + 2y_1 + y_2 + 2y_3 + y_4).
$$

Also,

$$
\begin{aligned}
Z_3(u) &\equiv Z(u) \bmod (u + 2) \\
&= z_0 + z_1 + z_2 + z_3, \\
Y_3(u) &\equiv Y(u) \bmod (u + 2) \\
&= y_0 + y_1 + y_2 + y_3 + y_4.
\end{aligned}
$$

Let
$$
m_2 = (z_0 + z_1 + z_2 + z_3) \, (y_0 + y_1 + y_2 + y_3 + y_4).
$$

Also,

$$
\begin{aligned}
Z_4(u) &\equiv Z(u) \bmod (u^2 + u + 2) \\
&= (z_0 + z_2 + 2z_3) + (z_1 + 2z_2 + 2z_3) \, u, \\
Y_4(u) &\equiv Y(u) \bmod (u^2 + u + 2) \\
&= (y_0 + y_2 + 2y_3 + 2y_4) + (y_1 + 2y_2 + 2y_3) \, u.
\end{aligned}
$$

Let

$$
\begin{aligned}
m_3 &= (z_0 + z_2 + 2z_3)(y_0 + y_2 + 2y_3 + 2y_4), \\
m_4 &= (z_1 + 2z_2 + 2z_3)(y_1 + 2y_2 + 2y_3), \\
m_5 &= (z_0 + z_1 + z_3)(y_0 + y_1 + y_3 + 2y_4).
\end{aligned}
$$

Also,

$$
\begin{aligned}
Z_5(u) &\equiv Z(u) \bmod (u^2 + 2u + 2) \\
&= (z_0 + z_2 + z_3) + (z_1 + z_2 + 2z_3)\, u, \\
Y_5(u) &\equiv Y(u) \bmod (u^2 + 2u + 2) \\
&= (y_0 + y_2 + y_3 + 2y_4) + (y_1 + y_2 + 2y_3)\, u.
\end{aligned}
$$

Let

$$
\begin{aligned}
m_6 &= (z_0 + z_2 + z_3)(y_0 + y_2 + y_3 + 2y_4), \\
m_7 &= (z_1 + z_2 + 2z_3)(y_1 + y_2 + 2y_3), \\
m_8 &= (z_0 + z_1 + 2z_2)(y_0 + y_1 + 2y_2 + 2y_4).
\end{aligned}
$$

Let $m_9 = z_3 y_4$ (wraparound). The multiplications $m_0, m_1, \ldots, m_9$ are sufficient to compute $\Phi(u) = Z(u)Y(u)$.

The generator matrix of the associated (10, 4, 5) linear code over $GF(3)$ is given by

$$
C = \begin{bmatrix}
1 & 1 & 1 & 1 & 0 & 1 & 1 & 0 & 1 & 0 \\
0 & 2 & 1 & 0 & 1 & 1 & 0 & 1 & 1 & 0 \\
0 & 1 & 1 & 1 & 2 & 0 & 1 & 1 & 2 & 0 \\
0 & 2 & 1 & 2 & 2 & 1 & 1 & 2 & 0 & 1
\end{bmatrix}.
$$

The above code compares well with the only known nontrivial *perfect* nonbinary (11,6,5) ternary code [5.2]. Also, it can easily be seen that the length of ternary codes is smaller than the length of binary codes for the same values of k and d. This is due to the fact that as the field of constants grows in size, the number of polynomials of any degree defined over the field also grows.

Table 5.2 contains a list of selected ternary codes that can be obtained from the CRT-based aperiodic convolution algorithms.

Table 5.2: Selected Ternary Linear Codes Obtained From the Aperiodic Convolution Algorithm.

Length of Conv. N	(n, k, d) Parameters
5	$(6,3, 3)$
6	$(7, 4, 3)$, $(7, 2, 5)$
7	$(9, 5, 3)$, $(9, 3, 5)$
8	$(10, 6, 3)$, $(10, 4, 5)$, $(10, 2, 7)$
9	$(12, 7, 3)$, $(12, 5, 5)$, $(12, 3, 7)$
10	$(13, 8, 3)$, $(13, 6, 5)$, $(13, 4, 7)$
11	$(15, 9, 3)$, $(15, 7, 5)$, $(15, 5, 7)$, $(15, 3, 9)$
12	$(17, 10, 3)$, $(17, 8, 5)$, $(17, 6, 7)$, $(17, 4, 9)$

5.2.3   A Shift Register-Based Encoding Procedure

Let $P(u)$ be of the form $P(u) = \prod_{i=1}^{L} P_i(u)$ and let $Z(u)$ represent the length k information vector as a polynomial of degree $k - 1$. The encoding procedure corresponds to the following steps:

(1) Compute $Z_i(u) \equiv Z(u) \bmod P_i(u)$, $i = 1, 2, \ldots, L$.

(2) Form appropriate linear combinations of the coefficients of $Z_i(u)$, so that the product $Z_i(u)Y_i(u) \bmod P_i(u)$ can be computed.

(3) Form appropriate linear combinations of the coefficients of $\hat{Z}(u)$, so that the product $\hat{Z}_i(u)\hat{Y}_i(u) \bmod u^s$ can be computed (wraparound).

Note that step (3) can be treated as a special case of step (2) and, therefore, we will only study steps (1) and (2) for the encoder design.

It is clear that $Z_i(u)$ is the remainder obtained by dividing $Z(u)$ by $P_i(u)$. Such a procedure can be implemented by a division circuit, which is a σ_i-stage shift register with feedback connections determined

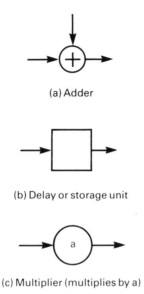

(a) Adder

(b) Delay or storage unit

(c) Multiplier (multiplies by a)

Figure 5.1: The Three Basic Units for Shift Register Implementation.

according to $P_i(u)$ $(\sigma_i = \deg[P_i(u)])$ [5.2]. Figure 5.1 shows the three basic types of units that constitute a feedback shift register. Based on Figure 5.1, the general configuration of a shift register-based division circuit is shown in Figure 5.2.

The multipliers and adders shown in Figure 5.2 are for the appropriate field. In the binary case, the multipliers and adders are for $GF(2)$ and, consequently, each of the multiplier coefficients is either 0 (open circuit) or 1 (short circuit) and the adders are replaced by exclusive-OR gates. We require L such circuits, one for each of the polynomials $P_i(u)$. Figure 5.3 (a)-(f) shows the shift register configurations with feedback connections for some of the polynomials defined over $GF(2)$.

If $P_i(u)$ is irreducible over the field of constants, $Z_i(u)Y_i(u)$ mod $P_i(u)$ is computed by first computing $Z_i(u)Y_i(u)$ and then reducing the product modulo $P_i(u)$. In this case, the equations for the linear combinations of the coefficients of $Z_i(u)$ are independent of $P_i(u)$. Figure 5.4 shows the linear combinations of the coefficients of $Z_i(u)$ as required by step (2) for some of the polynomials $P_i(u)$ defined over $GF(2)$.

We now consider the encoder design for the (24, 6, 9) binary code. Here $P(u) = u^2(u^2+1)\ (u^2+u+1)\ (u^3+u^2+1)\ (u^3+u+1)$ and $s = 2$.

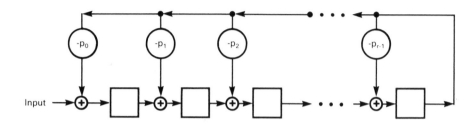

Figure 5.2: A Shift Register Circuit for Division by a Polynomial $P(u) = u^r + p_{r-1}u^{r-1} + \cdots + p_1 u + p_0$.

The shift register implementation of the encoder is shown in Figure 5.5. The information vector is represented as $Z(u) = z_0 + z_1 u + \cdots + z_5 u^5$. The encoder circuit requires 14 delay units and 17 exclusive-OR gates.

5.3 Error Detection

With this technique of error control, a received block of digits corresponding to a transmitted code word is inspected to ascertain if it is a valid code word. An error pattern will go undetected by the decoder *iff* it is identical to one of the nonzero code words. This certainly is the case if the all-zero code word is transmitted, but also holds for the transmission of an arbitrary code word, due to the linearity of the code [5.5].

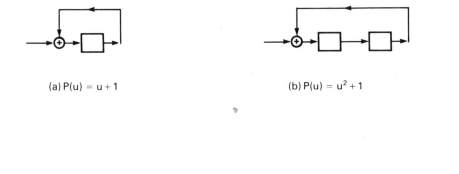

(a) P(u) = u + 1

(b) P(u) = u² + 1

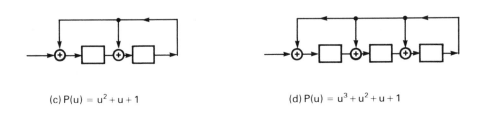

(c) P(u) = u² + u + 1

(d) P(u) = u³ + u² + u + 1

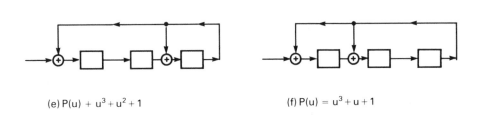

(e) P(u) + u³ + u² + 1

(f) P(u) = u³ + u + 1

Figure 5.3: Examples of Shift Register Circuits for Division by a Polynomial $P(u)$.

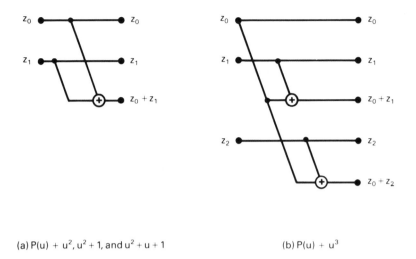

(a) $P(u) + u^2, u^2 + 1$, and $u^2 + u + 1$ (b) $P(u) + u^3$

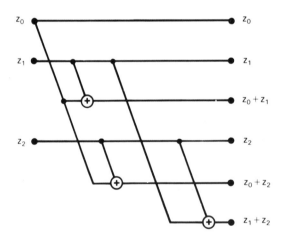

(c) $P(u) + u^3 + u + 1$ and $u^3 + u^2 + 1$

Figure 5.4: Linear Combinations of the Coefficients of $Z(u)$ Required for the Computation $Z(u)Y(u) \bmod P(u)$.

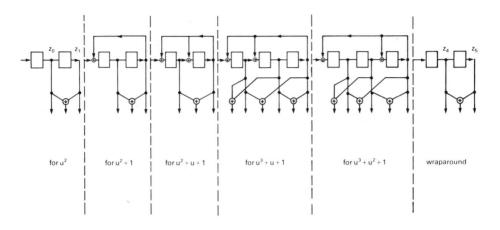

Figure 5.5: A Shift Register Implementation of the Encoder for the (24, 6, 9) Binary Code.

An (n, k) linear code forms a vector space of dimension k, with corresponding null space of dimension $(n-k)$. The null space is spanned by a set of $(n-k)$ $\ell.i.$ vectors; hence, a received vector is assumed error-free if it satisfies $(n - k)$ $\ell.i.$ equations. In the following, we establish a procedure to obtain the $(n - k)$ $\ell.i.$ equations for a given family of codes of length, or complexity, n.

We see from Section 5.2 that the generator matrix C of such a linear code has a *block structure*. Such a structure arises due to a computation of the type

$$\Phi_i(u) \equiv Z_i(u)Y_i(u) \bmod P_i(u),$$

for each of the L relatively prime polynomials $P_i(u)$ of degree σ_i, $i = 1, 2, \ldots, L$. Also, there is a last block that corresponds to the computation $\hat{Z}(u)\hat{Y}(u) \bmod u^s$ (wraparound). There are $(L + 1)$ blocks in the generator matrix C of the (n, k, d) code; we label these blocks of C as C_i, $i = 1, 2, \ldots, L+1$. The number of columns in each block is given by M_{σ_i}, $i = 1, 2, \ldots, L + 1$. Since *each one* of the blocks corresponds to a computation of the type described above, it is clear that in each block C_i, there are exactly σ_i $(\sigma_{t+1} = s)$ columns which are $\ell.i.$ and the remaining $[M_{\sigma_i} - \sigma_i]$ columns are linearly dependent on them. Each such block gives rise, therefore, to $[M_{\sigma_i} - \sigma_i]$ parity-check equations. It is assumed that the dimension k of the code under investigation is greater than σ_i, $i = 1, 2, \ldots, L + 1$. This assumption does not restrict our analysis but rather simplifies it. Any received vector can be segmented into $L + 1$ blocks, and the part of the received vector that corresponds to each block C_i should satisfy the $[M_{\sigma_i} - \sigma_i]$ parity-check equations for that block. A conventional procedure may be employed to find the $[M_{\sigma_i} - \sigma_i]$ $\ell.i.$ parity-check equations on each block, with the reader referred to [5.2] for details. We shall, however, illustrate the technique in the sequel by an example.

Thus, we have established $[M_{\sigma_i} - \sigma_i]$ parity-check equations satisfied by the code digits corresponding to each one of the blocks. Moreover, these equations are also $\ell.i.$ of the $[M_{\sigma_j} - \sigma_j]$ parity-check equations satisfied by the code symbols corresponding to the block C_j, $j \neq i$. The total number of parity-check equations obtained by this straightforward procedure is

$$
\begin{aligned}
\sum_{i=1}^{L+1} [M_{\sigma_i} - \sigma_i] &= \sum_{i=1}^{L+1} M_{\sigma_i} - \sum_{i=1}^{L+1} \sigma_i \\
&= n - N \\
&= (n - k) - (d - 1),
\end{aligned}
$$

since $N = k + d - 1$.

This set of $(n - N)$ parity-check equations does not change when the value of k is increased (decreased), as n and N are constant for a given set of linear codes obtained from the convolution algorithm for a fixed length. We also observe that there is a need to obtain an additional $(d - 1)$ parity-check equations which are $\ell.i.$ of those obtained above. These are obtained from the relations *between* the columns of different blocks of C.

Each block $C_i, i = 1, 2, \ldots, L$, of C consists of two parts. The first σ_i columns of each block can always be arranged so as to correspond to the polynomial

$$
Z_i(u) \equiv Z(u) \bmod P_i(u), i = 1, 2, \ldots, L.
$$

The remaining $[M_{\sigma_i} - \sigma_i]$ columns arise due to the multiplicative complexity of the block C_i; we have already established the parity-check equations corresponding to these. In each block C_i, let us consider only the first σ_i columns. The polynomial $Z(u)$ can be recovered from the residue polynomials $Z_i(u)$, $i = 1, 2, \ldots, L$, using the CRT. Such a reconstruction gives a polynomial of degree $D - 1$, where D is the degree of the polynomial $P(u)$. Since the information polynomial is of degree $k - 1$, however, for a reconstructed polynomial to correspond to an information polynomial, its last $[(D - 1) - (k - 1)] = (D - k)$ coefficients should be identically zero. If $Z(u)$ is the reconstructed polynomial

1. The first k coefficients are the information bits if no errors have taken place.

2. The expressions for the last $(D - k)$ coefficients provide us with $(D - k)$ $\ell.i.$ parity-check equations, i.e., if the reconstructed polynomial $Z(u)$ is

$$Z(u) = z_0 + z_1 u + \cdots + z_{D-1} u^{D-1};$$

then we have

$$z_i = 0, \quad i = k, k + 1, \ldots, D - 1. \tag{5.4}$$

The first σ_{L+1} $(= s)$ columns of the block corresponding to the wraparound provide us with the polynomial

$$\bar{Z}(u) \equiv \hat{Z}(u) \bmod u^s = \bar{z}_{k-1} + \bar{z}_{k-2} u + \cdots + \bar{z}_{k-s} u^{s-1}. \tag{5.5}$$

If no errors have taken place, then the corresponding coefficients of the polynomial $Z(u)$ and $\bar{Z}(u)$ should be the same, i.e.,

$$z_i = \bar{z}_i, \quad i = k - 1, k - 2, \ldots, k - s. \tag{5.6}$$

The relation (5.6) results in an additional set of s $\ell.i.$ equations. The total number of equations given by (5.4) and (5.6) is $(D - k) + s = (D + s - k)$. Also, $N = (D + s) = (k + d - 1)$ and, therefore, $(D + s - k) = (d - 1)$. The parity-check equations given by (5.4) and (5.6) provide the remaining $(d - 1)$ parity-check equations. These equations are also $\ell.i.$, due to the fact that any coefficient of a polynomial cannot be expressed as a linear sum of the other coefficients.

Of these $(d - 1)$ parity-check equations, obtained from (5.4) and (5.6), the parity-check equations given by (5.6) are altered and $d_2 = |d_1 - d|$ equations are added to (5.4) or dropped from (5.4) when a new code is selected from the same set, depending on whether k is decreased or increased, respectively, where d_1 is the design minimum distance of the selected code. We see that all the codes in a given *family* have the same code length n. It is also worthwhile noting that the above procedure not only gives the required set of parity-check equations but also provides the equations that can be used to find the transmitted message polynomial once the received code vector is found to contain no errors. The complete procedure to find the parity-check equations is illustrated below for the $(12, 4, 5)$ code.

EXAMPLE 4. Consider the $(12, 4, 5)$ linear code over $GF(2)$. Since $N = k + d - 1 = 8$, we have $k + d = 9$. The polynomials are $Z(u) = \sum_{i=0}^{3} z_i u^i$ and $Y(u) = \sum_{i=0}^{4} y_i u^i$. We choose $P(u) = u^2 (u^2 + 1)(u^2 + u + 1)$ and $s = 2$ to compute the aperiodic convolution $\Phi(u) = Z(u)Y(u)$. Let $P_1(u) = u^2$, $P_2(u) = (u^2 + 1)$, and $P_3(u) = (u^2 + u + 1)$. Reducing the polynomials $Z(u)$ and $Y(u)$ modulo each of $P_i(u)$, we obtain

$$
\begin{aligned}
Z_1(u) &\equiv Z(u) \bmod u^2 \\
&= z_0 + z_1 u, \\
Y_1(u) &\equiv Y(u) \bmod u^2 \\
&= y_0 + y_1 u.
\end{aligned}
$$

Let

$$
\begin{aligned}
m_0 &= z_0 y_0, \\
m_1 &= z_1 y_1, \\
m_2 &= (z_0 + z_1)(y_0 + y_1).
\end{aligned}
$$

Also,

$$
\begin{aligned}
Z_2(u) &\equiv Z(u) \bmod (u^2 + 1) \\
&= (z_0 + z_2) + (z_1 + z_3)u, \\
Y_2(u) &\equiv Y(u) \bmod (u^2 + 1) \\
&= (y_0 + y_2 + y_4) + (y_1 + y_3)u.
\end{aligned}
$$

Let

$$
\begin{aligned}
m_3 &= (z_0 + z_2)(y_0 + y_2 + y_4), \\
m_4 &= (z_1 + z_3)(y_1 + y_3), \\
m_5 &= (z_0 + z_1 + z_2 + z_3)(y_0 + y_1 + y_2 + y_3 + y_4).
\end{aligned}
$$

Also,

$$
\begin{aligned}
Z_3(u) &\equiv Z(u) \bmod (u^2 + u + 1) \\
&= (z_0 + z_2 + z_3) + (z_1 + z_2)u, \\
Y_3(u) &\equiv Y(u) \bmod (u^2 + u + 1) \\
&= (y_0 + y_2 + y_3) + (y_1 + y_2 + y_4)u.
\end{aligned}
$$

Let

$$
\begin{aligned}
m_6 &= (z_0 + z_2 + z_3)\,(y_0 + y_2 + y_3), \\
m_7 &= (z_1 + z_2)\,(y_1 + y_2 + y_4), \\
m_8 &= (z_0 + z_1 + z_3)\,(y_0 + y_1 + y_3 + y_4).
\end{aligned}
$$

For the wraparound, let

$$
\begin{aligned}
m_9 &= z_3 y_4, \\
m_{10} &= z_2 y_3, \\
m_{11} &= (z_3 + z_2)\,(y_4 + y_3).
\end{aligned}
$$

It is easy to establish that the above 12 multiplications are sufficient to determine the polynomial products $\Phi_i(u) \equiv Z_i(u)Y_i(u) \bmod P_i(u)$, as well as to compute the polynomial product $Z(u)Y(u)$.

The generator matrix of the corresponding (12, 4, 5) code is given by

$$
C = \left[
\begin{array}{ccc|ccc|ccc|ccc}
1 & 0 & 1 & 1 & 0 & 1 & 1 & 0 & 1 & 0 & 0 & 0 \\
0 & 1 & 1 & 0 & 1 & 1 & 0 & 1 & 1 & 0 & 0 & 0 \\
0 & 0 & 0 & 1 & 0 & 1 & 1 & 1 & 0 & 0 & 1 & 1 \\
0 & 0 & 0 & 0 & 1 & 1 & 1 & 0 & 1 & 1 & 0 & 1 \\
\multicolumn{3}{c|}{\text{for}} & \multicolumn{3}{c|}{\text{for}} & \multicolumn{3}{c|}{\text{for}} & \multicolumn{3}{c}{\text{wrap-}} \\
\multicolumn{3}{c|}{P_1(u)} & \multicolumn{3}{c|}{P_2(u)} & \multicolumn{3}{c|}{P_3(u)} & \multicolumn{3}{c}{\text{around}}
\end{array}
\right]
$$

$$
= [C_1|C_2|C_3|C_4].
$$

We mention that the design minimum distance of the above code is 5, while the actual minimum distance is 6. Each of the blocks C_1, C_2, and C_3 corresponding to the computations modulo $P_1(u)$, $P_2(u)$, and $P_3(u)$, respectively, provide $M_2 - 2 = 3 - 2 = 1$ parity-check equation. These equations are given by

$$\begin{array}{llll}
\text{for } P_1(u) & : & c_0 + c_1 + c_2 = 0, & \\
\text{for } P_2(u) & : & c_3 + c_4 + c_5 = 0, & \\
\text{for } P_3(u) & : & c_6 + c_7 + c_8 = 0, & (5.7)
\end{array}$$

where c_i is the ith digit of the code vector $\mathbf{c} = [c_0 c_1 \cdots c_{11}]^T$. The block C_4 for the wraparound computation provides a single equation, given by

$$c_9 + c_{10} + c_{11} = 0. \tag{5.8}$$

Now, if we take

$$Z_i(u) \equiv Z(u) \bmod P_i(u), \quad i = 1, 2, 3,$$

then, using the CRT, the polynomial $Z(u)$ can be written as

$$Z(u) = z_0 + z_1 u + \cdots + z_5 u^5,$$

where

$$\begin{array}{rcl}
z_0 & = & c_0, \\
z_1 & = & c_1, \\
z_2 & = & c_0 + c_1 + c_4 + c_6, \\
z_3 & = & c_0 + c_3 + c_4 + c_7, \\
z_4 & = & c_1 + c_3 + c_4 + c_6, \\
z_5 & = & c_0 + c_1 + c_3 + c_7.
\end{array}$$

If $Z(u)$ corresponds to a message polynomial of degree 3, then,

$$z_4 = z_5 = 0. \tag{5.9}$$

Also, the wraparound corresponds to the polynomial coefficients z_2 and z_3 and, therefore,

$$z_3 + c_9 = 0 \text{ or } c_0 + c_3 + c_4 + c_7 + c_9 = 0,$$

and

$$z_2 + c_{10} = 0 \text{ or } c_0 + c_1 + c_4 + c_6 + c_{10} = 0. \tag{5.10}$$

The equations (5.7) and (5.8), along with (5.9) and (5.10), provide the complete set of $n - k = 12 - 4 = 8$ *l.i.* parity-check equations that a received vector must satisfy before it can be declared error-free for the (12, 4, 6) code.

If we were now to write the parity-check equations for the (12, 6, 3) code, (5.7) and (5.8) remain unchanged. For $k = 6$, the message polynomial is of degree 5 and, therefore, there is no parity-check equation corresponding to (5.9). The parity-check equations due to wraparound become $z_5 + c_9 = 0$ and $z_4 + c_{10} = 0$. Thus, only two equations are altered when the parity-check equations for the (12, 6, 3) code are obtained from the parity-check equations for the (12, 4, 6) code.

5.3.1 Burst Error Detection Capability

In this section, a procedure is outlined which can be used to determine the burst error detection capability of the codes under investigation. The information vector can be expressed as a polynomial $Z(u)$ of degree $(k - 1)$. Consider the polynomial $Z_i(u)$ obtained as

$$Z_i(u) \equiv Z(u) \bmod P_i(u).$$

Clearly, $Z_i(u)$ is zero *iff* $Z(u)$ is a multiple of $P_i(u)$. Consequently, if $Z(u)$ is a multiple of $P_i(u)$, then the block corresponding to the computation

$$\Phi_i(u) \equiv Z_i(u)Y_i(u) \bmod P_i(u) \tag{5.11}$$

is zero.

Let the blocks corresponding to C_{i_1}, C_{i_2} be zero in a code vector. Since $Z(u)$ is a degree $(k-1)$ polynomial, the following constraint must be satisfied for the blocks C_{i_1}, $C_{i_2}, \ldots,$

$$\sum_{i=i_1,i_2\ldots} \sigma_i \leq k - 1.$$

In addition, we also have

$$\sum_{i=1}^{L+1} \sigma_i = k + d - 1.$$

Combining the above two expressions, we obtain

$$\sum_{\substack{i=1 \\ i \neq i_1, i_2, \ldots}}^{L+1} \sigma_i \geq d.$$

This discussion is summarized in the following lemma.

LEMMA 5.1 *For any arbitrary information vector, the sum of the degrees of the polynomials $P_i(u)$ such that the associated blocks of the code vector are nonzero, is at least d. Conversely, if there is a vector having nonzero elements corresponding to C_{i_1}, $C_{i_2}, \ldots$, and*

$$\sigma_{i_1} + \sigma_{i_2} + \cdots < d, \tag{5.12}$$

then, such a vector cannot be a code vector.

We further assume that the computation in (5.11) is performed so that each of the blocks C_i is an $(n_i, \sigma_i, \sigma_i)$ linear code, where $n_i = M_{\sigma_i}$ (a mathematical basis and further explanation for this assumption is provided in the next section). For every code vector, the number of nonzero elements in a nonzero block is, therefore, at least σ_i.

Lemma 5.1 and the above assumption can be used to determine the burst error detection capability of the codes. For example, for the (24, 6, 9) code, if blocks $C_1, C_2, \ldots, C_6$ correspond to $P_1(u) = u^2$, $P_2(u) = (u^2+1)$, $P_3(u) = (u^3+u+1)$, $P_4(u) = (u^2+u+1)$, $P_5(u) = (u^3 + u^2 + 1)$, and the wraparound $s = 2$, then it can be shown that the code can detect error bursts of length up to 12. Unfortunately, the burst error detection capability for these codes depends on the ordering of the blocks and, therefore, it is not possible to derive a closed-form expression, as is true in the case of cyclic codes [5.2].

5.4 Error Correction

In this section, we derive a procedure for performing error correction for the linear codes obtained from the aperiodic convolution algorithms. We again consider the block structure of the generator matrix. The polynomial $P(u)$ factors into L relatively prime polynomials

$P_i(u)$, $1, 2, \ldots, L$, and there are L such blocks in the generator matrix of the code, each block corresponding to a computation of the type

$$\Phi_i(u) \equiv Z_i(u)Y_i(u) \bmod P_i(u).$$

The last block of the generator matrix arises due to the computation of the wraparound coefficients of the ordinary polynomial product $\Phi(u)$. A key property of such a block structure is described in the following lemma.

LEMMA 5.2 *If the polynomial $P_i(u)$ is irreducible, then the block C_i corresponding to the computation $\Phi_i(u) \equiv Z_i(u)Y_i(u) \bmod P_i(u)$ is an $(n_i, \sigma_i, \sigma_i)$ linear code, where $n_i = M_{\sigma_i}$.*

PROOF. For all the polynomials $P_i(u)$ that are irreducible, the computation $\Phi_i(u) \equiv Z_i(u)Y_i(u) \bmod P_i(u)$ is carried out in two steps:

(1) Compute $\Phi_i'(u) = Z_i(u)Y_i(u)$.

(2) Reduce $\Phi_i(u) \equiv \Phi_i'(u)$ modulo $P_i(u)$.

Step (1) corresponds to the computation of the ordinary polynomial product; therefore, if $\deg[P_i(u)] = \sigma_i$, then such a computation generates an $(n_i, \sigma_i, \sigma_i)$ code, where $n_i = M_{\sigma_i}$, the multiplicative complexity associated with the algorithm for step (1).

The only form of polynomials for which the above procedure is *not* adopted are the polynomials $P_i(u) = u^{\sigma_i}$, $P_i(u) = (u + a_i)^{\sigma_i}$, $a_i \epsilon \mathcal{F}$, and the computation of the wraparound coefficients which is given by the computation $\hat{Z}(u)\hat{Y}(u) \bmod u^s$. In our analysis of the decoding procedure, however, we assume that computations of the type $Z_i(u)Y_i(u) \bmod (u + a_i)^{\sigma_i}$, $a_i \epsilon \mathcal{F}$, are also performed using steps (1) and (2) of Lemma 5.2. It may be observed that this assumption does not result in a higher complexity algorithm (which also means a code of longer length for the same dimension k and minimum distance d) for values of $\sigma_i \leq 2$. For values of $\sigma_i > 2$, the rise in multiplicative complexity is very marginal. Also, the actual rise in complexity depends on the field used for the computation. For example, in the binary case, for $\sigma_i = 3$ and 4, the complexity only increases by one multiplication.

Since each block C_i, $i = 1, 2, \ldots, L + 1$, is associated with a computation of the type given in step (1) of Lemma 5.2, each block is an $(n_i, \sigma_i, \sigma_i)$ code, $i = 1, 2, \ldots, L+1$, where $n_i = M_{\sigma_i}$ and $\sigma_{L+1} = s$. We, therefore, start our attempt to decode a received vector by first partitioning the received digits into $L + 1$ blocks, and then *independently* decoding the received digits for each block C_i. If $ZD_i(u)$ represents the polynomial corresponding to the decoded vector for the block C_i, then $ZD_i(u)$ is the same as the transmitted residue $Z_i(u)$ for the block C_i, if no more than $\lfloor (\sigma_i - 1)/2 \rfloor$ errors take place in the part of the received vector corresponding to C_i.

The residue polynomial $Z_i(u)$ is decoded erroneously only if more than $\lfloor (\sigma_i - 1)/2 \rfloor$ or at least $(\frac{\sigma_i}{2} + 1)$ (for even σ_i) or at least $(\frac{\sigma_i - 1}{2} + 1)$ (for odd σ_i) errors occur in the block C_i. Note that for σ_i even, a decoding failure takes place if the received digits for the block C_i contain $\sigma_i/2$ errors. If a decoding failure takes place in a block, then such a block can be eliminated from further analysis. Elimination of such a block C_i essentially means that we have to analyze a code of dimension k and of minimum distance $d - \sigma_i$. We have, however, also eliminated *at least* $\lfloor (\sigma_i - 1)/2 \rfloor + 1$ errors by excluding such a block. If we recover $Z(u)$ from the reduced code (code obtained by eliminating a block), we can still correct a maximum of $\lfloor (d - 1)/2 \rfloor$ errors in the overall received vector.

Let α_i be the number of errors in that part of the received vector corresponding to the block C_i. We have to establish a procedure to recover $Z(u)$ from the received vector if

$$\sum_{i=1}^{L+1} \alpha_i \leq \lfloor (d - 1)/2 \rfloor.$$

Once the decoding for each block C_i is performed, there are two possibilities, namely

1. It is error free, i.e., $\alpha_i \leq \lfloor (\sigma_i - 1)/2 \rfloor$;

2. It is incorrectly decoded, i.e., $\alpha_i > \lfloor (\sigma_i - 1)/2 \rfloor$.

Now, let the blocks C_{i_1}, $C_{i_2}, \ldots, C_{i_f}$ be decoded *incorrectly*. The least number of errors in the received vector for such an event to take place is given by

$$\min\{\sum_{j=i_1,i_2,\ldots,i_f} \alpha_j\} = \sum_{j=i_1,i_2,\ldots,i_f} \{\lfloor(\sigma_j-1)/2\rfloor + 1\}.$$

The decoder is to correct the errors for such an event only if

$$\lfloor(d-1)/2\rfloor \geq \sum_{j=i_1,i_2,\ldots,i_f} \{\lfloor(\sigma_j-1)/2\rfloor + 1\}.$$

It is easy to show from the above that

$$d > \sum_{j=i_1,i_2,\ldots,i_f} \sigma_j,$$

or, alternatively,

$$\sum_{\substack{j=1 \\ j\neq i_1,i_2,\ldots,i_f}}^{L+1} \sigma_j \geq k. \tag{5.13}$$

The above discussion can be summarized in the form of a theorem.

THEOREM 5.1 *If the number of errors that take place in a code vector is at most the error correcting capability of the code, $\lfloor(d-1)/2\rfloor$, then after each block C_i is decoded according to its minimum distance σ_i, there is at least one set of blocks which is error-free such that the sum of the degrees of the polynomials $P_j(u)$ corresponding to these blocks is at least k.*

We will use this theorem for further analysis. Since the information vector is represented as a polynomial $Z(u)$ of degree $(k-1)$, it can be recovered using the CRT from any set of residues of the type

$$Z_\ell(u) \equiv Z(u) \bmod P_\ell(u), \quad \ell = \ell_1, \ell_2, \ldots,$$

provided that,

$$\sum_{\ell=\ell_1,\ell_2,\ldots} \sigma_\ell \geq k.$$

Let π be the set of integers $\pi = \{1, 2, \ldots, L+1\}$. The integer i in the set π corresponds to the polynomial $P_i(u)$ of degree σ_i. The integer

$L + 1$ corresponds to the wraparound. From this integer set, we form subsets of integers $\pi_1, \pi_2, \ldots$ such that each set is the *minimal* set with respect to the property that the sum of the degrees of the polynomials corresponding to the integers in each subset is at least k. Let this sum be k_1 for the subset π_1 and so on. Since $k_1 \geq k$, we can use the CRT to construct a polynomial $Z_1'(u)$ of degree $(k_1 - 1)$ using the residue polynomials $Z_\ell(u)$, $\ell \epsilon \pi_1$. This procedure can also be performed for the subsets $\pi_2, \pi_3, \ldots$ and so on. If the constructed polynomial $Z_1'(u)$ for the subset π_1 is of degree $(k_1 - 1)$, then, clearly, such a polynomial has to satisfy the following $(k_1 - k)$ equations for it to be accepted as a *candidate* for the transmitted polynomial:

$$z_{1i}' = 0, \quad i = k, k+1, \ldots, k_1 - 1, \tag{5.14}$$

where

$$Z_1'(u) = z_{10}' + z_{11}'u + z_{12}'u^2 + \cdots + z_{1,k_1-1}'u^{k_1-1}.$$

Similar relationships hold for the remaining subsets $\pi_2, \pi_3 \ldots$.

Let $Z_1'(u)$, $Z_2'(u), \ldots$ be the *candidates* for the transmitted message polynomial. As the code under consideration has a minimum distance d, the code vector corresponding to a *valid* information polynomial will differ from the received vector in a maximum of $\lfloor (d-1)/2 \rfloor$ places. Theorem 5.1 guarantees the *existence* of at least one candidate information polynomial which is the valid information polynomial, provided the number of errors present in the received vector is at most the error-correcting capability of the code. Hence, if the code vector corresponding to a candidate information polynomial differs from the received vector in at most $\lfloor (d-1)/2 \rfloor$ places, it is accepted as the valid information polynomial.

The complete decoding algorithm for the codes may be enumerated as follows:

Step 1: Partition the received vector according to blocks $C_1, C_2, \ldots,$ C_{L+1} of C.

Step 2: Perform the decoding for each block independently.

Step 3: Discard the blocks for which a decoding failure takes place.

Step 4: Using the CRT, construct the candidates for the information polynomial from the residue polynomials obtained from the blocks declared error-free in **Step 3**.

Step 5: Construct the candidate code vectors for each of the candidate information polynomials.

Step 6: Accept a candidate code vector as a valid code vector if it differs from the received vector in at most $\lfloor (d-1)/2 \rfloor$ places. The corresponding candidate information polynomial is then accepted as a valid information polynomial.

The Chinese Remainder Theorem establishes the *uniqueness* of the decoding procedure described above. A block diagram of the decoder is given in Figure 5.6. The decoding algorithm for each of the blocks is the *same* as the decoding algorithm described above, as these blocks are obtained from a computation of the same nature as that of the original code. Since each of the blocks is a code of small dimension and minimum distance, as compared to the dimension and minimum distance of the original code, we can examine these codes further to simplify the design of the decoder for these blocks. For example, the small (3,2,2), (6,3,3), and (9,4,4) codes are one-step *majority logic decodable*, and this fact can be incorporated into the design of the decoder for the blocks. Clearly, the decoder configuration is such that it can be implemented using a *parallel architecture*, a feature which may be useful in high data rate communication systems. Furthermore, since codes of this type form the basis of the overall decoding procedure, it is *plausible* that the overall performance of the system may be improved by using *soft-decision* decoding for the small codes [5.6] as opposed to the above *hard-decision* approach. This issue will be pursued further in Chapter 6.

Let us briefly analyze the complexity of the other blocks constituting the decoder. Consider the implementation for the reconstruction corresponding to the set π_1. The set π_1 corresponds to the CRT reconstruction of the polynomial $Z_1'(u)$ of degree $(k_1 - 1)$ from the knowledge of the residue polynomials $Z_\ell(u) \equiv Z(u) \bmod P_\ell(u)$, $\ell \epsilon \pi_1$. Using the CRT, $Z_1'(u)$ is obtained as

$$Z_1'(u) \equiv \sum_{\ell \epsilon \pi_1} [S_\ell(u) Z_\ell(u)] \bmod P_1'(u), \qquad (5.15)$$

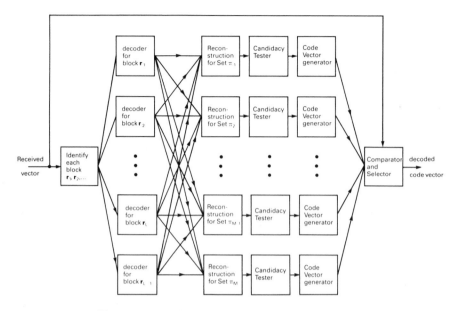

Figure 5.6: Block Diagram of the Decoder.

where

$$P_1'(u) = \prod_{\ell \epsilon \pi_1} P_\ell(u).$$

The polynomials $S_\ell(u)$ are determined *a priori* from knowledge of the $P_\ell(u)$, $\ell \epsilon \pi_1$. The expression (5.15) can be implemented in three steps as follows:

(1) Compute $S_\ell(u) Z_\ell(u)$, $\ell \epsilon \pi_1$.

(2) Form the sum $\sum_{\ell \epsilon \pi_1} S_\ell(u) Z_\ell(u)$.

(3) Reduce the resulting polynomial modulo $P_1'(u)$.

Step (1) can be implemented by a multiplication circuit as shown in Figure 5.7 [5.2]. Step (2) is implemented using a series of adders, one for each coefficient. Step (3) is implemented by a feedback shift register configuration as shown in Figure 5.2. The reconstruction for the sets π_2, π_3, ... can be implemented in a similar manner. A shift register-based implementation of the code vector generator was discussed in

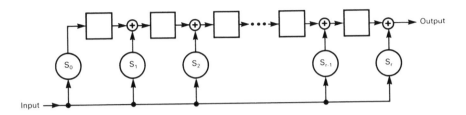

Figure 5.7: A Shift-Register Circuit for Multiplication by a Polynomial
$S(u) = s_r u^r + s_{r-1} u^{r-1} + \cdots + s_1 u + s_0$.

Section 5.2.3. Finally, the implementation of the candidacy tester and
the comparator and selector blocks is clear from the description of the
decoding algorithm.

Also consider the decoder design when k is increased (decreased),
keeping the length n of the code constant. **Steps 1, 2,** and **3** of the
decoding algorithm are unaltered. The decoder circuit for **Step 4** of
the algorithm has to be suitably modified. It has been shown in Section
5.2.1 that a change in k does not imply a significant change in the encoding procedure and, therefore, **Step 5** does not involve a significant
change as k is varied. In **Step 6**, we only have to adjust the threshold
of the comparator to the error correcting capability, $\lfloor (d-1)/2 \rfloor$, of the
new code. In the following, we present two examples to illustrate the
decoding algorithm developed above.

EXAMPLE 5. Consider the (8,4,3) code. $P(u)$ is chosen as $P(u) =
u(u^2+1)(u^2+u+1)$ and $s = 1$. The received vector is partitioned into
four blocks C_1, C_2, C_3, and C_4 corresponding to $P_1(u) = u$, $P_2(u) =
(u^2+1)$, $P_3(u) = (u^2+u+1)$, and the wraparound, respectively. The
blocks C_1 and C_4 are (1,1,1) codes for which no decoder is required.
The decoder for blocks C_2 and C_3 is a decoder for the (3,2,2) code and
simply checks the block for even parity. If this test fails, the block is
rejected. The set π is given by

$$\pi = \{1, 2, 3, 4\},$$

with possible subsets

$$\pi_1 = \{1, 2, 4\}, \ \pi_2 = \{2, 3\}, \text{ and } \pi_3 = \{1, 3, 4\}.$$

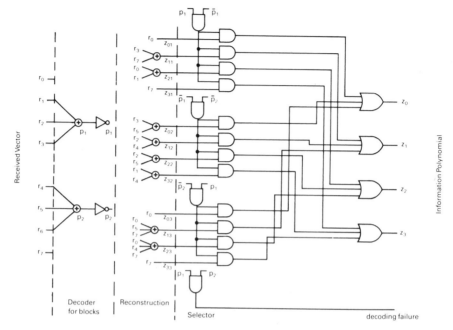

Figure 5.8: Decoder Circuit for the (8,4,3) Binary Code.

The subsets π_1, π_2, and π_3 correspond to the different choices of the residue polynomials that can be used to recover the information polynomial. No candidacy tester and comparator are required for the subsets. If we are interested only in the recovery of the information polynomial, the code vector generator is also not required in this case. A circuit diagram for the resulting decoder is given in Figure 5.8. The generator matrix of the code is derived in Section 5.2.1.

EXAMPLE 6.   Consider the (24,6,9) code. The polynomials are $P_1(u) = u^2$, $P_2(u) = (u^2 + u + 1)$, $P_3(u) = (u^2 + 1)$, $P_4(u) = (u^3 + u + 1)$, $P_5(u) = (u^3 + u^2 + 1)$, and $s = 2$. The received vector is to be partitioned into 6 blocks C_1, C_2, ..., C_6 corresponding to the polynomials $P_1(u)$, $P_2(u)$, ..., $P_5(u)$, and the wraparound, respectively. The blocks C_1, C_2, C_3, and C_6 are (3,2,2) codes and the blocks C_4 and C_5 are (6,3,3) codes. The decoders for the blocks C_1, C_2, C_3, and C_6 are, therefore, simple parity-checks and a block is rejected if it has odd parity. The decoders for blocks C_4 and C_5 are identical. This decoder is a one-step majority logic decoder. The set π is

$$\pi = \{1, 2, 3, 4, 5, 6\}$$

and there are 17 possible subsets: $\pi_1 = \{4, 5\}$, $\pi_2 = \{1, 2, 3\}$, $\pi_3 = \{2, 3, 6\}$, $\pi_4 = \{1, 2, 6\}$, $\pi_5 = \{1, 3, 6\}$, $\pi_6 = \{1, 2, 4\}$, $\pi_7 = \{1, 3, 4\}$, $\pi_8 = \{1, 4, 6\}$, $\pi_9 = \{2, 3, 4\}$, $\pi_{10} = \{2, 4, 6\}$, $\pi_{11} = \{3, 4, 6\}$, $\pi_{12} = \{1, 2, 5\}$, $\pi_{13} = \{1, 3, 5\}$, $\pi_{14} = \{1, 5, 6\}$, $\pi_{15} = \{2, 3, 5\}$, $\pi_{16} = \{2, 5, 6\}$, and $\pi_{17} = \{3, 5, 6\}$.

For a *parallel implementation*, the decoder requires 17 blocks for reconstruction of the polynomials to be tested for candidacy. The candidacy tester is present only for the subsets π_6, $\pi_7, \ldots$, π_{17}. The code vector generator is not enabled if the candidacy tester fails. The comparator is a set of 24 exclusive-OR gates followed by a counter and threshold detector set at 5.

Given the decoder for the (24,6,9) code, we analyze the overhead required to decode the (24,8,7) code. The design of the decoder for the blocks C_1, $C_2, \ldots$, C_6 is unaltered. There are 13 subsets for the (24,8,7) code: $\pi_1 = \{1, 2, 3, 6\}$, $\pi_2 = \{1, 2, 3, 4\}$, $\pi_3 = \{2, 3, 4, 6\}$, $\pi_4 = \{1, 2, 4, 6\}$, $\pi_5 = \{1, 3, 4, 6\}$, $\pi_6 = \{1, 2, 3, 5\}$, $\pi_7 = \{2, 3, 5, 6\}$, $\pi_8 = \{1, 2, 5, 6\}$, $\pi_9 = \{1, 3, 5, 6\}$, $\pi_{10} = \{1, 4, 5\}$, $\pi_{11} = \{2, 4, 5\}$, $\pi_{12} = \{3, 4, 5\}$, $\pi_{13} = \{4, 5, 6\}$, and we require 13 reconstruction blocks. The subsets π_1, π_{10}, π_{11}, π_{12}, and π_{13} do not require a candidacy test. The design of the code vector generator and the comparator is essentially as before, except that the threshold detector is now set at 4.

5.4.1 Experimental Results

Since majority logic decoding is a simple method of determining the error digits from the parity-check sums, the majority logic decodability of a larger number of linear codes belonging to this new class of linear codes has been examined. The reader is referred to [5.2, 5.3] for a detailed description of majority logic decoding of linear codes.

At the receiving end, we are generally interested in reliably recovering only the information symbols, therefore, we restrict our attention to finding orthogonal check sums on the error digits that correspond to the message digits. The following definition characterizes the majority logic decodability of linear codes.

DEFINITION 1.   A linear code with minimum distance d is said to be *sufficiently orthogonalizable in one step iff* it is possible to form $J = d - 1$ parity check sums orthogonal on each of the error digits that correspond to the message digits.

For a code that is sufficiently orthogonalizable, we can estimate the message vector correctly even if $\lfloor (d - 1)/2 \rfloor$ errors take place in the received vector. We also note that a completely orthogonalizable code is always sufficiently orthogonalizable, but the converse may not be true. In our analysis, the following codes were found to be sufficiently orthogonalizable in one step:

(6,3,3)	1 error-correcting
(8,4,3)	1 error-correcting
(10,5,3)	1 error-correcting
(10,3,5)	2 error-correcting
(12,6,3)	1 error-correcting
(12,4,5)	2 error-correcting
(14,3,7)	3 error-correcting
(14,5,5)	2 error-correcting
(16,4,7)	3 error-correcting
(16,6,5)	2 error-correcting
(18,5,7)	3 error-correcting
(18,3,9)	4 error-correcting
(20,4,9)	4 error-correcting
(22,5,9)	4 error-correcting
(22,3,11)	5 error-correcting
(24,4,11)	5 error-correcting
(26,5,11)	5 error-correcting
(30,5,13)	6 error-correcting

Also, the (24,6,9) code was found to be a *two-step* sufficiently orthogonalizable code. The parity-check equations for the above listed codes are given in Appendix A.

5.4.2 Code Family Encoding/Decoding Relationships

It has been shown that the encoding/decoding methods utilize the block structure of the codes arising due to the computation $Z_i(u)Y_i(u)$ modulo the relatively prime factor polynomials $P_i(u)$, $i = 1, 2, \ldots L$, and the wraparound of s points. Such a structure can be used to advantage to design codes with the *same* dimension k but *different* minimum distance d. In this section, we analyze the block structure of the generator matrix and show that it is possible to obtain a group of codes of dimension k, having different error correcting capability, in such a way that the encoder/decoder design for this group of codes is essentially the same. Such codes find application in communication systems that employ *variable redundancy codes* for adaptive error correction [5.3].

For the computation of an aperiodic convolution of length $N = k + d - 1$, the degree D of the polynomial $P(u)$ and the number of wraparound points s satisfy the relation

$$N = k + d - 1 = D + s. \tag{5.16}$$

If the polynomial $P(u)$ has L relatively prime polynomial factors, $P_i(u)$, each of degree σ_i, then

$$D = \sum_{i=1}^{L} \sigma_i. \tag{5.17}$$

Combining (5.16) and (5.17), we obtain

$$k + d - 1 = \sum_{i=1}^{L+1} \sigma_i, \tag{5.18}$$

where $\sigma_{L+1} = s$. Based on (5.18), we state the following result:

LEMMA 5.3 *If the block C_j corresponding to the computation of the polynomial product $\Phi_j(u) \equiv Z_j(u)Y_j(u) \mod P_j(u)$ is deleted from C, the resulting matrix is the generator matrix of an $(n - n_j, k, d - \sigma_j)$ code.*

PROOF. The expression (5.18) can be written in the form

$$k + (d - \sum_{j=i_1, i_2, \ldots} \sigma_j) - 1 = \sum_{\substack{i=1 \\ i \neq i_1, i_2, \ldots}}^{L+1} \sigma_i. \tag{5.19}$$

The length of the code is equal to the sum total of the multiplicative complexity associated with each block; that is,

$$n = \sum_{i=1}^{L+1} M_{\sigma_i}.$$

In accordance with (5.19), the length of the code can be written as

$$n - \sum_{j=i_1,i_2,\ldots} M_{\sigma_j} = \sum_{\substack{i=1 \\ i\neq i_1,i_2,\ldots}}^{L+1} M_{\sigma_i}. \qquad (5.20)$$

The expression (5.19) indicates that it is possible to obtain a code of dimension k and minimum distance d' from a given code of dimension k and minimum distance $d(d' < d)$ by discarding the computation with regard to the factor polynomials $P_j(u)$, $j = i_1, i_2, \ldots$, such that

$$d' = d - \sum_{j=i_1,i_2,\ldots} \sigma_j.$$

At the same time, (5.20) indicates that the generator matrix of such a code is obtained by simply dropping the columns of the original generator matrix belonging to the blocks C_j, $j = i_1, i_2, \ldots$, and also establishes the length n' of the new code.

Using Lemma 5.3, the generator matrix C' of the (n', k, d') code is obtained from the generator matrix C of the (n, k, d) code by appropriately deleting columns of C. For a given value of d', the various σ_j, $j = i_1, i_2, \ldots$, are chosen so as to minimize n'. Such a choice then leads to the most efficient code possible for this procedure.

Let us consider the (24,6,9) code to illustrate this unique feature of these linear codes. The various polynomials are $P_1(u) = u^2$, $P_2(u) = (u^2+u+1)$, $P_3(u) = (u^2+1)$, $P_4(u) = (u^3+u+1)$, $P_5(u) = (u^3+u^2+1)$, and $s = 2$. The generator matrix for this code has 6 blocks. From the (24,6,9) code, we can obtain the generated matrix of the (21,6,7) code by dropping any one of the 4 blocks (having 3 columns each) C_1, C_2, C_3, or C_6 from the generator matrix of the (24,6,9) code. Using similar techniques, we can obtain the generator matrices for the (18,6,6), (15,6,4), and (12,6,3) codes by discarding suitable groups of columns from the generator matrix C of the (24,6,9) code.

Given the decoder of an (n, k, d) code, we now consider the problem of designing a decoder for the (n', k, d') code obtained by dropping the blocks $\{C_J\} = \{C_j : j = i_1, i_2, \ldots\}$ from the generator matrix C of the (n, k, d) code. The blocks present in the resulting (n', k, d') code are also present in the original (n, k, d) code and, therefore, **Steps 1, 2,** and **3** of the decoding procedure described earlier are *unaltered.* The decoders for the blocks $\{C_J\}$ are simply disabled. In **Step 4** of the procedure, the reconstruction for a subset π_β is disabled if the subset π_β has any of the decoder outputs corresponding to the blocks $\{C_J\}$ as input. The reconstruction for the remaining subsets is the same as that of the (n, k, d) code. In **Step 5**, the outputs of the code vector generator corresponding to the blocks $\{C_J\}$ are again disabled, so as to obtain the code vectors of length n'. In **Step 6**, the threshold of the comparator is now set at $\lfloor (d' - 1)/2 \rfloor$, the error correcting capability of the new code.

Since the (n', k, d') code is a lower minimum distance code as compared to the (n, k, d) code, it is expected that the decoder structure for the (n', k, d') code would be simpler than that for the (n, k, d) code, which is indeed the case. It is clear from the above description that the decoder for the (n', k, d') code is obtained by simply disabling a portion of the decoder for the (n, k, d) code. As a result, the *processing throughput* of the decoder for all such codes is the same for a parallel implementation.

For example, consider the decoder design for the (21,6,7) code obtained from the (24,6,9) code by dropping the block C_1. We have $\{C_J\} = \{C_1\}$ and, consequently, the decoder for the block C_1 is disabled in **Step 1**. In **Step 4**, the reconstruction for the subsets π_2, π_4, π_5, π_6, π_7, π_8, π_{12}, π_{13}, and π_{14} are disabled, as all these subsets have the output of the decoder for the block C_1 as one of their inputs. Note that this also reduces the number of comparisons to be performed at the final stage of the decoding procedure. For the (21,6,7) code, the threshold of the comparator is reduced to 4 in **Step 6**.

A similar approach can be adopted to obtain an (n', k, d') code from an (n, k, d) code when $d' > d$. In this case, groups of columns are appended to the generator matrix of the (n, k, d) code. Every new group of columns corresponds to a computation with respect to a new polynomial $P_j(u)$. Care must be taken, however, to ensure that the polynomial $P_j(u)$ be *coprime* to all the existing polynomials $P_i(u)$, $i = 1, 2, \ldots, L$; $j \neq i$. The decoder structure for such a code can be easily analyzed. It is worthwhile to observe that the decoder would be more complex in design and require additional hardware; however, the additional processing requirements can easily be determined due to the *highly structured, systematic* decoding procedure shown in Figure 5.6.

5.5   Discussion

As described in Chapters 3 and 5, it is clear from the construction of bilinear algorithms for aperiodic (or periodic) convolution that the associated multiplicative complexity is directly dependent upon the field selected, the complexities of small degree polynomial multiplication algorithms, and the choice of the modulo polynomial $P(u)$. A number of algorithms are possible, and in this work effort is directed toward the construction of algorithms which possess as low a multiplicative complexity as possible. The formulation of the problem is such that the additive complexity of the algorithms does not play a significant role in the design of the algorithms. The examples presented in Chapters 3 and 5 illustrate that a wide variety of algorithms may be generated.

It is of interest to compare the actual complexity of the aperiodic convolution algorithm to its theoretical lower bound. If the field of constants, $\mathcal{F}$, has at least N elements, where $N = k + d - 1$, then a length N convolution may be computed in N multiplications [5.7]. The corresponding code is a *maximum-distance-separable* (n, k, d) code, where $n = k + d - 1$ [5.2]. The only nontrivial $(k \neq 1,\ d \neq 1)$ maximum-distance-separable code defined over $GF(2)$ is the $(3,2,2)$ code. On the other hand, the product of two polynomials $Z(u)$ and $Y(u)$ having degrees $k - 1$ and $d - 1$, respectively, can be computed in kd multiplications in a straightforward manner. The corresponding code is a (kd, k, d) code, which is a trivial extension of the $(d, 1, d)$ *repetition code* (in the (kd, k, d) code, each of the k information bits is repeated d times). We are not familiar with any meaningful bound on the multiplicative complexity of aperiodic convolution algorithms (such a bound must depend on the number of distinct elements in $\mathcal{F}$) other than the two extreme cases mentioned above.

If $P_i(u) = Q_i^{\beta_i}(u)$, where the polynomial $Q_i(u)$ is irreducible over $\mathcal{F}$, then the minimum number of multiplications required to compute $\Phi_i(u) \equiv Z_i(u)Y_i(u) \bmod P_i(u)$ is $2\sigma_i - 1$, where $\deg[P_i(u)] = \sigma_i$ [5.8]. This result can be extended to obtain a lower bound on the multiplicative complexity of the CRT-based procedure to compute the aperiodic convolution of length N. If $P(u)$ has L factors and a wraparound of $s(s \geq 1)$ coefficients is used to compute $\Phi(u) = Z(u)Y(u)$, then the lower bound on the multiplicative complexity of the procedure (and the length of the associated code) is $\mu = 2N - (L + 1)$. For example, for $N = 14$, $P(u) = u^2(u^2 + u + 1)(u^2 + 1)(u^3 + u + 1)(u^3 + u^2 + 1)$, $s = 2$, and, therefore, $\mu = 28 - 6 = 22$. This compares favorably with the actual multiplicative complexity of the aperiodic convolution algorithm of length 14, which is 24.

Several lower bounds are available in the coding theory literature on the value of n for given values of k and d for an (n, k, d) code over $\mathcal{F}$ [5.1,5.2]. The bounds are also applicable to the multiplicative complexity n of the algorithm for computing the aperiodic convolution of two sequences of lengths k and d. For example, it is not possible to compute the aperiodic convolution of two sequences of lengths $k = 8$ and $d = 5$ in less than 16 multiplications over $GF(2)$. There is another way in which a lower bound on the multiplicative complexity of aperiodic convolution algorithms can be obtained. We now discuss the relevant ideas.

Thus far, we have analyzed the structure and properties of the NC system of expressions represented by $C(A\mathbf{x} \bullet B\mathbf{y})$ for the system of bilinear forms Ψ in (2.15) and its A-dual Φ in (2.22). Let us consider the *B-dual* of Ψ given by the system of expressions $B^T(A\mathbf{x} \bullet C^T\mathbf{z})$. From the indeterminate commuted A-dual, $A^T(B\mathbf{y} \bullet C^T\mathbf{z})$, it can be established that the B-dual corresponds to the system Δ of d bilinear forms:

$$\Delta = X^T\mathbf{z} = \begin{bmatrix} x_0 & x_1 & x_2 & \cdots & x_{k-2} & x_{k-1} \\ x_1 & x_2 & x_3 & \cdots & x_{k-1} & x_k \\ x_2 & x_3 & x_4 & \cdots & \vdots & \vdots \\ \vdots & \vdots & \vdots & & & \\ x_{d-1} & x_d & x_{d+1} & \cdots & x_{k+d-3} & x_{k+d-2} \end{bmatrix} \begin{bmatrix} z_0 \\ z_1 \\ \vdots \\ z_{k-1} \end{bmatrix}.$$

Comparison of (2.15) with the above bilinear form reveals that since $\Delta = B^T(A\mathbf{x} \bullet C^T\mathbf{z})$, the $(d \times n)$-dimensional matrix B^T generates a linear $(n, d, \bar{k})$ code over $\mathcal{F}$, where $\bar{k} \geq k$. As a result, if $C(A\mathbf{x} \bullet B\mathbf{y})$ is a computation of the system of bilinear forms Φ in (2.15), then C is the generator matrix of an $(n, k, \bar{d})$ code $(\bar{d} \geq d)$ and B^T is the generator matrix of an $(n, d, \bar{k})$ code $(\bar{k} \geq k)$. This is an inherent limitation of the linear codes obtained from the aperiodic convolution approach. For example, it is not possible to obtain the $(k+1, k, 2)$ binary code, $k > 2$, (it is a single parity-check code) using this approach, as it would also imply the existence of a $(k + 1, 2, k)$ binary code. This discussion does lead, however, to a useful *necessary* condition for the existence of an aperiodic convolution algorithm of two sequences of lengths k and d. This condition is stated as a lemma.

LEMMA 5.4 *The aperiodic convolution of two sequences of lengths k and d can be computed in n multiplications over $\mathcal{F}$, only if there exist $(n, k, \bar{d})$ and $(n, d, \bar{k})$ linear codes over $\mathcal{F}$, $\bar{d} \geq d$, $\bar{k} \geq k$.*

For example, based on the lemma, we can say that there does not exist an algorithm over $GF(2)$ for aperiodic convolution of two sequences having lengths 5 and 9 that requires less than 20 multiplications. Note that there do not exist $(n, 5, 9)$ and $(n, 9, 5)$ binary linear codes for which $n < 20$. It is worthwhile mentioning that the CRT-based aperiodic convolution algorithm gives rise to (22,5,9) and (22,9,5) binary linear codes (refer to Table 5.1). A question of fundamental importance arises at this stage: "If there exist $(n, k, \bar{d})$ and $(n, d, \bar{k})$ linear codes over $\mathcal{F}$, $\bar{d} \geq d$, and $\bar{k} \geq k$, does this imply the existence of an aperiodic convolution algorithm over $\mathcal{F}$ requiring only n multiplications?" This is an open research problem.

The connection between the CRT and linear codes was first studied by Stone [5.9] and subsequently by Bossen et al [5.10] and Mandelbaum [5.11]. These codes are referred to as *redundant residue codes* in [5.1]. The algebraic construction and properties of redundant residue codes are different from the linear codes obtained here. The CRT is employed to obtain the error-correcting properties of the redundant residue codes in [5.9-5.11], while in this work it is employed as a means for computing the aperiodic convolution.

It has been shown in Chapter 3 that a bilinear algorithm over $GF(p)$ which is valid for input data over $GF(p)$ is also valid for input data over $GF(p^m)$; therefore, the results of this chapter, including the technique to design the bilinear algorithm, remain valid even if the prime 2 is replaced everywhere by a power of 2. This increases the field of constants, a fact which can be incorporated in the design of the algorithm in order to reduce the multiplicative complexity and, hence, the length of the codes over $GF(2^m)$.

The linear codes obtained from the algorithms compare well with similar already known codes in terms of their rate and distance properties. The complexity of the decoding algorithm is in direct proportion to the number of relatively prime factors of the polynomial $P(u)$. A large number of such factors increases the complexity of the decoder while minimizing the overall length of the code for specified values of k and d. A tradeoff is, therefore, possible between the complexity of the decoder and the length of the code.

Another interesting feature of the decoding algorithm is that it uses small length codes as a basis for the decoding of a long length code. The design of the decoder for the codes of small length and distance is very simple as compared to the design for a long length code. Because of this *modular structure*, the decoder can also detect and correct several error patterns having Hamming weight greater than the error-correcting capability of the code. For example, the decoder circuit of the (8,4,3) code given in Figure 5.8 can detect 32 percent of all the possible error patterns of Hamming weight 2. A large number of these codes were also found to be *ℓ-step majority logic decodable*, with $\ell \geq 1$. A general statement regarding algorithm design and majority logic decodability eludes us at this time and is an open area of research.

Among the binary codes generated from the aperiodic convolution algorithms, a large number of these codes were analyzed with the help of the computer in order to determine the actual minimum distance of the code as compared to the design minimum distance. The actual minimum distance of most of the codes tested was found to be the *same* as the design minimum distance. The weight distribution for a selected number of codes is given in Appendix B.

5.6 References

[5.1] F. J. McWilliams and N. J. A. Sloane, *The Theory of Error Correcting Codes*, North Holland, 1977.

[5.2] W. W. Peterson and E. J. Weldon, Jr., *Error Correcting Codes*, MIT Press, 1978.

[5.3] S. Lin and D. J. Costello, Jr., *Error Control Coding: Fundamentals and Applications*, Prentice Hall, 1983.

[**5.4**] H. J. Helgert and R. D. Stinaff, "Minimum Distance Bounds for Binary Linear Codes," *IEEE Trans. Inform. Theory*, vol. IT-19, pp. 344-356, May 1973.

[**5.5**] J. K. Wolf, A. M. Michelson, and A. H. Levesque, "On the Probability of Undetected Error for Linear Block Codes," *IEEE Trans. Commun.*, vol. COM-30, pp. 317-324, Feb. 1982.

[**5.6**] G. C. Clark and J. B. Cain, *Error Correction Coding for Digital Communication*, Plenum Press, 1981.

[**5.7**] A. Lempel and S. Winograd, "A New Approach to Error Correcting Codes," *IEEE Trans. Inform. Theory*, vol. IT-23, pp. 503-508, July 1977.

[**5.8**] J. H. McClellan and C. M. Rader, *Number Theory in Digital Signal Processing*, Prentice Hall, 1979.

[**5.9**] J. J. Stone, "Multiple-Burst Error Correction with the Chinese Remainder Theorem," *Journ. Soc. Indust. Appl. Math.*, vol. 11, no. 1, pp. 74-81, 1963.

[**5.10**] D. C. Bossen and S. S. Yau, "Redundant Residue Polynomial Codes," *Inform. and Control*, vol. 13, pp. 597- 618, 1968.

[**5.11**] D. Mandelbaum, "A Method of Coding for Multiple Errors," *IEEE Trans. Inform. Theory*, vol. IT-14, pp. 518-521, 1968.

5.7 Appendix A

Orthogonal Parity-Check Equations for Sufficiently Orthogonalizable Codes

In this appendix, the orthogonal parity-check equations on each of the error digits $e_0, e_1, e_2, \ldots$, that correspond to the message digits are presented for some codes found to be sufficiently orthogonalizable. All the codes examined were one-step sufficiently orthogonalizable with the exception of the $(24,6,9)$ code which was found to be two-step sufficiently orthogonalizable.

1. The generator matrix of the $(6,3,3)$ code in systematic form is given by

$$C = \begin{bmatrix} 1 & 0 & 0 & 0 & 1 & 1 \\ 0 & 1 & 0 & 1 & 0 & 1 \\ 0 & 0 & 1 & 1 & 1 & 1 \end{bmatrix}.$$

The two parity-check sums orthogonal on e_0, e_1, and e_2 are

$$\{0 + 2 + 4, \ 0 + 3 + 5\}, \ \{1 + 2 + 3, \ 1 + 4 + 5\}, \ \text{and}$$

$$\{2 + 0 + 4, \ 2 + 1 + 3\},$$

respectively, where $\alpha + \beta + \gamma + \ldots$ corresponds to the parity-check sum $e_\alpha + e_\beta + e_\gamma + \ldots$.

2. The generator matrix of the $(8,4,3)$ code in systematic form is given by

$$C = \begin{bmatrix} 1 & 0 & 0 & 0 & 0 & 1 & 1 & 1 \\ 0 & 1 & 0 & 0 & 1 & 0 & 1 & 0 \\ 0 & 0 & 1 & 0 & 1 & 1 & 1 & 1 \\ 0 & 0 & 0 & 1 & 0 & 1 & 0 & 1 \end{bmatrix}.$$

The two parity-check sums orthogonal on e_0, e_1, e_2, and e_3 are

$$\{0 + 2 + 3 + 5, \ 0 + 4 + 6\}, \ \{1 + 2 + 4, \ 1 + 3 + 6 + 7\},$$

$$\{2 + 0 + 3 + 5, \ 2 + 1 + 4\}, \text{and} \ \{3 + 0 + 2 + 5, \ 3 + 1 + 6 + 7\},$$

respectively.

3. The generator matrix of the (10,5,3) code in systematic form is given by

$$C = \begin{bmatrix} 1 & 0 & 0 & 0 & 0 & 0 & 0 & 0 & 1 & 1 \\ 0 & 1 & 0 & 0 & 0 & 1 & 0 & 0 & 1 & 1 \\ 0 & 0 & 1 & 0 & 0 & 1 & 0 & 1 & 0 & 1 \\ 0 & 0 & 0 & 1 & 0 & 0 & 1 & 1 & 1 & 0 \\ 0 & 0 & 0 & 0 & 1 & 0 & 1 & 0 & 1 & 1 \end{bmatrix}.$$

The two parity-check sums orthogonal on e_0, e_1, e_2, e_3, and e_4 are

$$\{0+1+3+4+8,\ 0+2+5+6+7+9\},$$

$$\{1+2+5,\ 1+0+3+4+8\},\ \{2+1+5,\ 2+3+7\},$$

$$\{3+4+6,\ 3+2+7\},\ \text{and}\ \{4+3+6,\ 4+0+1+2+9\},$$

respectively.

4. The generator matrix of the (10,3,5) code in systematic form is given by

$$C = \begin{bmatrix} 1 & 0 & 0 & 0 & 0 & 1 & 1 & 1 & 0 & 1 \\ 0 & 1 & 0 & 1 & 0 & 1 & 1 & 1 & 1 & 0 \\ 0 & 0 & 1 & 1 & 1 & 1 & 0 & 0 & 1 & 1 \end{bmatrix}.$$

The four parity-check sums orthogonal on e_0, e_1, and e_2 are

$$\{0+3+5,\ 0+1+6,\ 0+4+7+8,\ 0+2+9\},$$

$$\{1+2+3,\ 1+5+9,\ 1+0+6,\ 1+4+8\},\ \text{and}$$

$$\{2+1+3,\ 2+4,\ 2+5+6,\ 2+0+7+8\},$$

respectively.

5. The generator matrix of the (12,6,3) code in systematic form is given by

$$C = \begin{bmatrix} 1 & 0 & 0 & 0 & 0 & 0 & 0 & 0 & 0 & 0 & 1 & 1 \\ 0 & 1 & 0 & 0 & 0 & 0 & 1 & 0 & 0 & 0 & 1 & 1 \\ 0 & 0 & 1 & 0 & 0 & 0 & 1 & 0 & 0 & 0 & 0 & 1 \\ 0 & 0 & 0 & 1 & 0 & 0 & 0 & 1 & 0 & 1 & 1 & 1 \\ 0 & 0 & 0 & 0 & 1 & 0 & 0 & 1 & 1 & 0 & 1 & 0 \\ 0 & 0 & 0 & 0 & 0 & 1 & 0 & 0 & 1 & 1 & 1 & 1 \end{bmatrix}.$$

The two orthogonal parity-check sums on $e_0, e_1, \ldots, e_5$ are

$$\{0+1+3+4+5+10, \ 0+6+7+8+11\}, \ \{1+2+6, \ 1+3+4+5+10\},$$

$$\{2+1+6, \ 2+4+10+11\}, \ \{3+4+7, \ 3+5+9\},$$

$$\{4+3+7, \ 4+5+8\}, \ \text{and} \ \{5+4+8, \ 5+3+9\},$$

respectively.

6. The generator matrix of the $(12,4,5)$ code in systematic form is given by

$$C = \begin{bmatrix} 1 & 0 & 0 & 0 & 0 & 0 & 0 & 1 & 0 & 1 & 1 & 1 \\ 0 & 1 & 0 & 0 & 0 & 1 & 1 & 0 & 0 & 1 & 1 & 1 \\ 0 & 0 & 1 & 0 & 1 & 0 & 1 & 1 & 1 & 0 & 1 & 0 \\ 0 & 0 & 0 & 1 & 1 & 1 & 0 & 0 & 1 & 0 & 1 & 1 \end{bmatrix}.$$

The four orthogonal parity-check sums on e_0, e_1, e_2, and e_3 are

$$\{0+2+7, \ 0+1+9, \ 0+3+6+10, \ 0+5+11\},$$

$$\{1+3+5, \ 1+2+6, \ 1+0+9, \ 1+4+7+11\},$$

$$\{2+3+4, \ 2+1+6, \ 2+0+7, \ 2+10+11\}, \ \text{and}$$

$$\{3+2+4, \ 3+1+5, \ 3+0+7+8, \ 3+9+11\},$$

respectively.

7. The generator matrix of the $(14,3,7)$ code in systematic form is given by

$$C = \begin{bmatrix} 1 & 0 & 0 & 0 & 0 & 1 & 1 & 1 & 0 & 1 & 1 & 1 & 0 & 1 \\ 0 & 1 & 0 & 1 & 0 & 1 & 1 & 0 & 1 & 0 & 1 & 1 & 0 & 0 \\ 0 & 0 & 1 & 1 & 1 & 1 & 0 & 0 & 0 & 1 & 0 & 1 & 1 & 1 \end{bmatrix}.$$

The six orthogonal parity-check sums on e_0, e_1, and e_2 are

$$\{0+6+8, \ 0+7, \ 0+2+9, \ 0+1+10, \ 0+3+11, \ 0+4+13\},$$

$$\{1+2+3, \ 1+5+9, \ 1+6+7, \ 1+8, \ 1+0+10, \ 1+11+13\},$$

and $\{2+1+3, \ 2+4, \ 2+0+9, \ 2+6+11, \ 2+12, \ 2+7+13\},$

respectively.

8. The generator matrix of the (14,5,5) code in systematic form is given by

$$
C = \begin{bmatrix}
1 & 0 & 0 & 0 & 0 & 1 & 0 & 1 & 0 & 1 & 1 & 1 & 1 & 1 \\
0 & 1 & 0 & 0 & 0 & 1 & 0 & 0 & 1 & 1 & 0 & 1 & 0 & 1 \\
0 & 0 & 1 & 0 & 0 & 0 & 1 & 1 & 0 & 1 & 0 & 0 & 1 & 1 \\
0 & 0 & 0 & 1 & 0 & 0 & 1 & 0 & 1 & 1 & 0 & 0 & 0 & 1 \\
0 & 0 & 0 & 0 & 1 & 0 & 0 & 0 & 0 & 0 & 1 & 1 & 1 & 1
\end{bmatrix}.
$$

The four orthogonal parity-check sums on $e_0, e_1, \ldots, e_4$ are

$$\{0+1+5,\ 0+2+7,\ 0+4+10,\ 0+3+6+8+9\},$$

$$\{1+3+8,\ 1+0+5,\ 1+10+11,\ 1+2+4+6+7+13\},$$

$$\{2+3+6,\ 2+0+7,\ 2+10+12,\ 2+1+4+5+8+13\},$$

$$\{3+2+6,\ 3+1+8,\ 3+0+5+7+9,\ 3+10+11+12+13\},$$

$$\text{and } \{4+0+10,\ 4+5+11,\ 4+7+12,\ 4+9+13\},$$

respectively.

9. The generator matrix of the (16,4,7) code in systematic form is given by

$$
C = \begin{bmatrix}
1 & 0 & 0 & 0 & 1 & 0 & 1 & 1 & 1 & 0 & 0 & 0 & 1 & 0 & 1 & 1 \\
0 & 1 & 0 & 0 & 1 & 1 & 0 & 0 & 1 & 1 & 1 & 0 & 1 & 1 & 1 & 0 \\
0 & 0 & 1 & 0 & 1 & 1 & 1 & 1 & 0 & 1 & 0 & 1 & 1 & 1 & 0 & 1 \\
0 & 0 & 0 & 1 & 1 & 0 & 1 & 0 & 1 & 1 & 1 & 0 & 0 & 1 & 0 & 1
\end{bmatrix}.
$$

The six orthogonal parity-check sums on e_0, e_1, e_2, and e_3 are

$$\{0+4+9,\ 0+2+7,\ 0+1+14,\ 0+8+10,\ 0+5+12,\ 0+3+11+15\},$$

$$\{1+2+5,\ 1+0+14,\ 1+3+10,\ 1+4+6,\ 1+8+11+15,\ 1+7+12\},$$

$$\{2+11,\ 2+0+7,\ 2+1+5,\ 2+4+8,\ 2+12+14,\ 2+10+13\}, \text{ and}$$

$$\{3+1+10,\ 3+4+12,\ 3+6+7,\ 3+8+14,\ 3+5+9,\ 3+0+2+15\},$$

respectively.

10. The generator matrix of the (16,6,5) code in systematic form is given by

$$C = \begin{bmatrix} 1 & 0 & 0 & 0 & 0 & 0 & 0 & 0 & 1 & 1 & 1 & 1 & 1 & 0 & 0 & 0 \\ 0 & 1 & 0 & 0 & 0 & 0 & 1 & 1 & 0 & 0 & 0 & 0 & 1 & 0 & 1 & 1 \\ 0 & 0 & 1 & 0 & 0 & 0 & 0 & 1 & 0 & 0 & 1 & 0 & 1 & 1 & 1 & 0 \\ 0 & 0 & 0 & 1 & 0 & 0 & 1 & 0 & 1 & 1 & 1 & 0 & 0 & 1 & 0 & 1 \\ 0 & 0 & 0 & 0 & 1 & 0 & 0 & 0 & 1 & 0 & 0 & 1 & 0 & 1 & 1 & 0 \\ 0 & 0 & 0 & 0 & 0 & 1 & 0 & 0 & 1 & 1 & 0 & 1 & 1 & 1 & 0 & 1 \end{bmatrix}.$$

The four orthogonal parity-check sums on e_0, $e_1, \ldots, e_5$ are

$$\{0 + 2 + 8 + 13, \ 0 + 1 + 9 + 15, \ 0 + 7 + 10, \ 0 + 4 + 5 + 11\},$$
$$\{1 + 3 + 6, \ 1 + 2 + 7, \ 1 + 0 + 9 + 15, \ 1 + 5 + 8 + 11 + 13 + 14\},$$
$$\{2 + 1 + 7, \ 2 + 3 + 4 + 10 + 11 + 15, \ 2 + 6 + 9 + 12, \ 2 + 0 + 8 + 13\},$$
$$\{3 + 1 + 16, \ 3 + 8 + 11, \ 3 + 4 + 9 + 12 + 14, \ 3 + 2 + 5 + 7 + 15\},$$
and $\{4 + 8 + 9, \ 4 + 7 + 14, \ 4 + 0 + 5 + 11, \ 4 + 2 + 1 + 13 + 15\},$

respectively.

11. The generator matrix of the (18,5,7) code in systematic form is given by

$$C = \begin{bmatrix} 1 & 0 & 0 & 0 & 0 \\ 0 & 1 & 0 & 0 & 0 \\ 0 & 0 & 1 & 0 & 0 \\ 0 & 0 & 0 & 1 & 0 \\ 0 & 0 & 0 & 0 & 1 \\ 0 & 1 & 0 & 1 & 0 \\ 1 & 0 & 1 & 1 & 0 \\ 0 & 0 & 1 & 1 & 0 \\ 1 & 1 & 1 & 0 & 0 \\ 0 & 1 & 0 & 1 & 1 \\ 1 & 0 & 1 & 1 & 1 \\ 0 & 0 & 1 & 0 & 1 \\ 0 & 1 & 1 & 0 & 0 \\ 1 & 1 & 0 & 1 & 1 \\ 0 & 1 & 0 & 0 & 1 \\ 1 & 0 & 1 & 1 & 1 \\ 1 & 1 & 1 & 1 & 0 \\ 0 & 1 & 1 & 1 & 1 \end{bmatrix}^T.$$

The six orthogonal parity-check sums on $e_0, e_1, \ldots, e_4$ are

$\{0+6+7, \; 0+8+12, \; 0+3+10+11, \; 0+9+13, \; 0+1+15+17,$

$0+2+5+16\}, \; \{1+3+5, \; 1+7+9+11, \; 1+2+12, \; 1+4+$

$14, 1+6+16, \; 1+0+15+17\}, \; \{0+6+13+14, \; 2+3+7, \; 2+4+11,$

$2+1+12, \; 2+0+5+16, \; 2+9+17\}, \; \{3+1+15, \; 3+2+7,$

$3+9+14, \; 3+0+11+15, \; 3+8+16, \; 3+4+12+17\}, \;$ and

$\{4+5+9, \; 4+6+10, \; 4+2+11, \; 4+1+14, \; 4+0+7+15,$

$4+3+12+17\},$

respectively.

12. The generator matrix of the (18,3,9) code in systematic form is given by

$$
C = \begin{bmatrix}
1 & 0 & 0 \\
0 & 1 & 0 \\
0 & 0 & 1 \\
1 & 1 & 0 \\
1 & 1 & 1 \\
1 & 0 & 0 \\
0 & 1 & 1 \\
1 & 0 & 1 \\
1 & 1 & 1 \\
0 & 0 & 1 \\
1 & 1 & 0 \\
1 & 0 & 0 \\
0 & 1 & 1 \\
1 & 0 & 1 \\
1 & 1 & 1 \\
1 & 1 & 0 \\
0 & 0 & 1 \\
1 & 0 & 1
\end{bmatrix}^T .
$$

The eight parity-check sums orthogonal on e_0, e_1, and e_2 are

$\{0+1+3, \; 0+4+6, \; 0+5, \; 0+2+7, \; 0+8+12, \; 0+11, \; 0+$

$13 + 16$, $0 + 9 + 17\}$, $\{1 + 0 + 3$, $1 + 4 + 7$, $1 + 2 + 6$, $1 + 8 + 13$, $1 + 5 + 10$, $1 + 12 + 16$, $1 + 14 + 17$, $1 + 11 + 15\}$, and $\{2 + 8 + 10$, $2 + 0 + 7$, $2 + 9$, $2 + 1 + 12$, $2 + 5 + 13$, $2 + 14 + 15$, $2 + 16$, $2 + 11 + 17\}$,

respectively.

13. The generator matrix of the $(20,4,9)$ code in systematic form is given by

$$
C =
\begin{bmatrix}
1 & 0 & 0 & 0 \\
0 & 1 & 0 & 0 \\
0 & 0 & 1 & 0 \\
0 & 0 & 0 & 1 \\
0 & 1 & 1 & 0 \\
1 & 1 & 1 & 0 \\
0 & 0 & 1 & 1 \\
1 & 1 & 1 & 1 \\
0 & 0 & 1 & 0 \\
1 & 1 & 0 & 1 \\
0 & 0 & 0 & 1 \\
0 & 1 & 1 & 1 \\
1 & 0 & 1 & 0 \\
0 & 1 & 1 & 0 \\
1 & 1 & 0 & 1 \\
1 & 0 & 1 & 1 \\
1 & 0 & 1 & 1 \\
1 & 1 & 1 & 0 \\
0 & 1 & 0 & 1 \\
1 & 0 & 0 & 1
\end{bmatrix}^{T}.
$$

The eight parity-check sums orthogonal on e_0, e_1, e_2, and e_3 are

$\{0 + 4 + 5$, $0 + 7 + 11$, $0 + 9 + 18$, $0 + 2 + 12$, $0 + 6 + 15$, $0 + 8 + 10 + 16$, $0 + 13 + 17$, $0 + 3 + 19\}$, $\{1 + 2 + 4$, $1 + 5 + 10 + 16$, $1 + 7 + 15$, $1 + 6 + 11$, $1 + 8 + 13$, $1 + 14 + 19$, $1 + 12 + 17$, $1 + 3 + 18\}$, $\{2 + 3 + 6$, $2 + 8$, $2 + 0 + 12$, $2 + 1 + 13$, $2 + 7 + 9$, $2 + 11$

$+18$, $2+15+19$, $2+10+14+17\}$, and $\{3+2+16$, $3+5+7$, $3+8+9+17$, $3+10$, $3+4+11$, $3+12+16$, $3+1+18$, $3+0+19\}$, respectively.

14. The generator matrix of the $(22,5,9)$ code in systematic form is given by

$$
C =
\begin{bmatrix}
1 & 0 & 0 & 0 & 0 \\
0 & 1 & 0 & 0 & 0 \\
0 & 0 & 1 & 0 & 0 \\
0 & 0 & 0 & 1 & 0 \\
0 & 0 & 0 & 0 & 1 \\
1 & 1 & 0 & 0 & 0 \\
1 & 0 & 1 & 0 & 0 \\
0 & 0 & 1 & 1 & 1 \\
0 & 0 & 1 & 1 & 0 \\
0 & 0 & 1 & 0 & 1 \\
1 & 0 & 1 & 1 & 0 \\
0 & 1 & 1 & 1 & 1 \\
1 & 0 & 0 & 1 & 1 \\
0 & 1 & 0 & 1 & 0 \\
1 & 1 & 0 & 0 & 1 \\
0 & 1 & 1 & 1 & 0 \\
0 & 0 & 0 & 1 & 1 \\
1 & 1 & 0 & 1 & 0 \\
0 & 1 & 1 & 0 & 1 \\
1 & 1 & 0 & 0 & 1 \\
1 & 0 & 1 & 0 & 0 \\
0 & 1 & 0 & 1 & 1 \\
\end{bmatrix}^T .
$$

The eight parity-check sums orthogonal on e_0, $e_1, \ldots, e_4$ are

$\{0+2+8$, $0+1+5$, $0+8+10$, $0+12+16$, $0+13+17$, $0+3+14+21$, $0+7+15+19$, $0+4+9+20\}$, $\{1+0+5$, $1+3+13$, $1+8+15$, $1+9+18$, $1+16+21$, $1+11+12+20$, $1+7+10+19$, $1+2+4+6+14\}$, $\{2+0+6$, $2+3+8$, $2+4+9$, $2+7+16$, $2+11+21$, $2+$

13+15, 2+1+5+14+18, 2+12+17+19+20}, {3+2+8, 3+1+13,

3+4+16, 3+7+9, 3+6+10, 3+11+18, 3+5+17, 3+0+14+

21}, and {4+2+9, 4+3+16, 4+7+8, 4+11+15, 4+5+14,

4+13+21, 4+0+1+18+20, 4+6+10+17+19},

respectively.

15. The generator matrix of the (22,3,11) code in systematic form is given by

$$
C = \begin{bmatrix}
1 & 0 & 0 \\
0 & 1 & 0 \\
0 & 0 & 1 \\
1 & 1 & 0 \\
1 & 0 & 1 \\
1 & 1 & 1 \\
1 & 0 & 1 \\
0 & 1 & 1 \\
1 & 1 & 0 \\
1 & 0 & 0 \\
0 & 1 & 0 \\
0 & 0 & 1 \\
1 & 1 & 0 \\
1 & 0 & 1 \\
0 & 1 & 1 \\
1 & 0 & 0 \\
0 & 1 & 0 \\
0 & 0 & 1 \\
1 & 1 & 0 \\
0 & 1 & 1 \\
1 & 0 & 1 \\
0 & 0 & 1
\end{bmatrix}^T .
$$

The ten parity-check sums orthogonal on e_0, e_1, and e_2 are

{0+1+3, 0+2+4, 0+5+7, 0+6+11, 0+8+10, 0+9, 0+12+16,

0+13+17, 0+15, 0+20+21}, {1+0+3, 1+4+5, 1+2+7, 1+

8+9, 1+10, 1+12+15, 1+11+14, 1+16, 1+17+18+20, 1+19+
21}, and $\{2+0+4,\ 2+3+5,\ 2+6+9,\ 2+7+10, 2+11,\ 2+$
$13+15,\ 2+1+14,\ 2+17,\ 2+16+19,\ 2+21\}$,

respectively.

16. The generator matrix for the (24,4,11) code in systematic form
    is given by

$$
C =
\begin{bmatrix}
1 & 0 & 0 & 0 \\
0 & 1 & 0 & 0 \\
0 & 0 & 1 & 0 \\
0 & 0 & 0 & 1 \\
1 & 0 & 1 & 0 \\
1 & 0 & 1 & 0 \\
1 & 1 & 0 & 0 \\
1 & 0 & 1 & 1 \\
1 & 1 & 1 & 1 \\
0 & 1 & 1 & 0 \\
1 & 0 & 0 & 1 \\
1 & 1 & 0 & 1 \\
0 & 0 & 0 & 1 \\
0 & 0 & 1 & 0 \\
1 & 1 & 0 & 0 \\
1 & 1 & 1 & 1 \\
0 & 0 & 1 & 1 \\
1 & 1 & 0 & 1 \\
0 & 1 & 0 & 0 \\
0 & 1 & 1 & 1 \\
1 & 0 & 0 & 1 \\
0 & 0 & 1 & 1 \\
1 & 0 & 1 & 0 \\
0 & 1 & 0 & 1
\end{bmatrix}^{T} .
$$

The ten parity-check sums orthogonal on e_0, e_1, e_2, and e_3 are

$\{0+2+4,\ 0+1+6,\ 0+3+10,\ 0+5+13,\ 0+7+16,\ 0$

$+8+19,\ 0+11+23,\ 0+12+20, 0+14+18,\ 0+9+17+$

21}, {1 + 0 + 6, 1 + 2 + 9, 1 + 3 + 23, 1 + 18, 1 + 7 + 8, 1+
11 + 20, 1 + 10 + 17, 1 + 19 + 21, 1 + 12 + 15 + 22, 1 + 5 + 13
+14}, {2 + 0 + 4, 2 + 1 + 9, 2 + 3 + 16, 2 + 13, 2 + 7 + 10, 2
+8 + 11, 2 + 12 + 21, 2 + 19 + 23, 2 + 15 + 17, 2 + 5 + 6 + 18},
and {3 + 0 + 10, 3 + 1 + 23, 3 + 2 + 16, 3 + 12, 3 + 4 + 7, 3+
6 + 11, 3 + 13 + 21, 3 + 14 + 17, 3 + 9 + 19, 3 + 5 + 8 + 18},
respectively.

17. The generator matrix for the (26,5,11) code in systematic form
is given by

$$
C = \begin{bmatrix}
1 & 0 & 0 & 0 & 0 \\
0 & 1 & 0 & 0 & 0 \\
0 & 0 & 1 & 0 & 0 \\
0 & 0 & 0 & 1 & 0 \\
0 & 0 & 0 & 0 & 1 \\
1 & 0 & 1 & 0 & 0 \\
1 & 1 & 0 & 0 & 0 \\
0 & 0 & 0 & 1 & 1 \\
1 & 1 & 1 & 1 & 0 \\
1 & 1 & 0 & 0 & 1 \\
0 & 0 & 1 & 1 & 1 \\
1 & 1 & 0 & 1 & 0 \\
1 & 0 & 1 & 1 & 1 \\
1 & 0 & 0 & 0 & 1 \\
0 & 1 & 1 & 0 & 1 \\
0 & 1 & 0 & 1 & 1 \\
0 & 0 & 1 & 1 & 0 \\
0 & 1 & 1 & 1 & 1 \\
1 & 1 & 1 & 0 & 1 \\
1 & 1 & 0 & 1 & 1 \\
1 & 0 & 0 & 1 & 0 \\
0 & 0 & 1 & 1 & 0 \\
1 & 0 & 1 & 0 & 0 \\
1 & 0 & 1 & 0 & 1 \\
0 & 1 & 0 & 1 & 0 \\
1 & 1 & 1 & 1 & 1
\end{bmatrix}^{T} .
$$

The ten parity-check sums orthogonal on e_0, $e_1, \ldots, e_4$ are

$\{0+2+5,\ 0+1+6,\ 0+3+20,\ 0+17+25,\ 0+4+13,\ 0+10+12,$

$0+11+24,\ 0+14+18,\ 0+15+19,\ 0+7+8+9+22\},\ \{1+$

$0+6,\ 1+3+24,\ 1+12+25,\ 1+18+23,\ 1+11+20,\ 1+9+13,$

$1+7+15,\ 1+4+17+21,\ 1+10+19+22,\ 1+2+5+8+16\},$

$\{2+0+5,\ 2+3+16,\ 2+19+25,\ 2+13+23,\ 2+9+18,\ 2+$

$15+17,\ 2+8+11,\ 2+7+10,\ 2+4+12+20,\ 2+1+21+24\},$

$\{3+0+20,\ 3+1+24,\ 3+2+16,\ 3+4+7,\ 3+6+11,\ 3+12+23,$

$3+14+17,\ 3+18+25,\ 3+5+10+13,\ 3+15+19+21+22\},$

and $\{4+0+13,\ 4+3+7,\ 4+8+25,\ 4+15+24,\ 4+22+23,$

$4+10+21,\ 4+11+19,\ 4+6+9,\ 4+1+16+17,\ 4+2+12+20\},$

respectively.

18. The generator matrix for the (30,5,13) code in systematic form is given by

$$
C = \begin{bmatrix}
1 & 0 & 0 & 0 & 0 \\
0 & 1 & 0 & 0 & 0 \\
0 & 0 & 1 & 0 & 0 \\
0 & 0 & 0 & 1 & 0 \\
0 & 0 & 0 & 0 & 1 \\
1 & 1 & 0 & 0 & 0 \\
1 & 0 & 1 & 0 & 0 \\
0 & 1 & 1 & 0 & 1 \\
0 & 0 & 0 & 1 & 1 \\
0 & 1 & 1 & 1 & 1 \\
1 & 1 & 1 & 0 & 1 \\
1 & 1 & 0 & 1 & 1 \\
0 & 0 & 1 & 1 & 0 \\
0 & 1 & 1 & 1 & 0 \\
1 & 1 & 1 & 1 & 1 \\
1 & 1 & 0 & 1 & 0 \\
1 & 0 & 0 & 0 & 1 \\
0 & 0 & 1 & 0 & 1 \\
1 & 0 & 0 & 1 & 0 \\
1 & 1 & 0 & 0 & 1 \\
1 & 0 & 1 & 1 & 0 \\
1 & 0 & 0 & 1 & 1 \\
0 & 1 & 1 & 1 & 1 \\
0 & 0 & 1 & 0 & 1 \\
0 & 1 & 0 & 1 & 0 \\
1 & 0 & 1 & 1 & 1 \\
1 & 1 & 1 & 1 & 0 \\
0 & 1 & 0 & 0 & 1 \\
1 & 0 & 1 & 0 & 1 \\
0 & 0 & 1 & 0 & 0
\end{bmatrix}^T .
$$

The twelve parity-check sums orthogonal on $e_0, e_1, \ldots, e_4$ are

$\{0 + 1 + 5,\ 0 + 2 + 6,\ 0 + 8 + 21,\ 0 + 18 + 29,\ 0 + 7 + 10,\ 0 + 9 + 14,\ 0 + 12 + 20,\ 0 + 13 + 26,\ 0 + 15 + 24,\ 0 + 4 + 16,\ 0 + 19 + 27,\ 0 + 11 + 22 + 25 + 28\},\ \{1 + 0 + 6,\ 1 + 11 + 21,\ 1 + 20 + 26,\ 1 + 7 + 17,\ 1$

+12+13, 1+2+8+9, 1+3+24, 1+14+25, 1+6+10+23+29, 1+

4+27, 1+16+19, 1+18+22+28}, {2+0+6, 2+7+27, 2+21+

25, 2+3+12, 2+10+19, 2+15+26, 2+4+17, 2+13+24, 2+

5+9+28, 2+29, 2+11+14, 2+8+16+20}, {3+4+8, 3+10+

14, 3+16+21, 3+2+12, 3+5+15, 3+13+23+27, 3+1+24, 3+

6+20, 3+17+18+28, 3+11+19, 3+7+9, 3+22+25+26+29},

and {4 + 3 + 8, 4 + 9 + 13, 4 + 11 + 15, 4 + 0 + 16, 4 + 16 + 26

4 + 5 + 6 + 7,   4 + 2 + 27, 4 + 5 + 19,   4 + 23 + 29, 4 + 1 + 27

4 + 20 + 25, 4 + 12 + 18 + 21},

respectively.

19. The generator matrix of the (24,6,9) code in systematic form is
    given by

$$C = \begin{bmatrix} 1 & 0 & 0 & 0 & 0 & 0 \\ 0 & 1 & 0 & 0 & 0 & 0 \\ 0 & 0 & 1 & 0 & 0 & 0 \\ 0 & 0 & 0 & 1 & 0 & 0 \\ 0 & 0 & 0 & 0 & 1 & 0 \\ 0 & 0 & 0 & 0 & 0 & 1 \\ 1 & 1 & 0 & 0 & 0 & 0 \\ 0 & 0 & 0 & 0 & 1 & 0 \\ 1 & 1 & 1 & 1 & 0 & 0 \\ 1 & 0 & 1 & 1 & 0 & 0 \\ 1 & 1 & 1 & 0 & 0 & 1 \\ 1 & 1 & 0 & 1 & 0 & 0 \\ 1 & 0 & 1 & 1 & 1 & 0 \\ 0 & 1 & 0 & 1 & 0 & 1 \\ 0 & 1 & 1 & 0 & 1 & 0 \\ 1 & 0 & 0 & 0 & 0 & 1 \\ 1 & 1 & 1 & 1 & 1 & 1 \\ 0 & 1 & 1 & 1 & 1 & 0 \\ 0 & 0 & 1 & 1 & 0 & 1 \\ 0 & 0 & 0 & 0 & 0 & 1 \\ 0 & 1 & 0 & 1 & 1 & 1 \\ 0 & 0 & 1 & 1 & 0 & 0 \\ 0 & 1 & 1 & 0 & 1 & 1 \\ 1 & 1 & 0 & 0 & 1 & 1 \end{bmatrix}^T .$$

Let $E_1^1 = e_0 + e_1$, $E_2^1 = e_0 + e_2$, $E_3^1 = e_0 + e_3$, $E_4^1 = e_0 + e_5$, $E_5^1 = e_0 + e_9$, $E_6^1 = e_0 + e_{11}$, $E_7^1 = e_0 + e_{12}$ and $E_8^1 = e_0 + e_{13}$ be eight selected sets of error digits. The eight parity-check sums orthogonal on E_1^1 are $\{4 + 5 + 13,\ 12 + 17,\ 3 + 11,\ 10 + 14 + 20,\ 9 + 13 + 18,\ 8 + 21,\ 6,\ 2 + 7 + 15 + 22\}$. Similarly, the eight parity-check sums orthogonal on E_2^1, E_3^1, E_4^1, E_5^1, E_6^1, E_7^1, and E_8^1 are

$$\{22 + 23,\ 16 + 20,\ 5 + 15 + 21,\ 3 + 4 + 12,\ 1 + 10 + 19,\ 9,\ 7 +$$

$$11 + 14,\ 6 + 13 + 18\},\ \{20 + 23,\ 16 + 22,\ 5 + 15,\ 2 + 4 + 12,\ 1 + 11,$$

$$9 + 21,\ 8 + 13 + 18,\ 7 + 10 + 14 + 19\},\ \{16 + 17,\ 3 + 15,\ 2 + 14 +$$

23, $4+11+20$, $1+10+21$, $9+18$, $7+8+22$, $6+13\}$, $\{2$, $3+$
21, $5+18$, $15+17+23$, $14+19+20$, $4+13+22$, $8+11$, $1+6+7$
$+12\}$, $\{1+3$, $4+5+20$, $13+19$, $12+18+23$, $10+15+21$, $8+9$,
$2+7+14$, $6+16+22\}$, $\{4+21$, $1+18+20$, $5+13+17$, $11+16$
$+19$, $10+23$, $8+9+14$, $2+7$, $6+15+22\}$, and $\{4+23$, $3+12+$
22, $11+19$, $10+21$, $1+9+18$, $8+14+20$, $2+7+16$, $5+6\}$,

respectively. From these orthogonal check sums, the sums E_1^1, E_2^1, $\dots$, E_8^1 can be correctly estimated, provided that there are no more than four errors in the error vector. Clearly, E_1^1, $E_2^1, \dots, E_8^1$ are orthogonal on e_0; hence, e_0 can be estimated from these sums. Furthermore, e_1, e_2, e_3, e_5, e_9, e_{11}, e_{12}, and e_{13} can be estimated from the above sums once e_0 is estimated.

Also, let $E_1^2 = e_4 + e_2$, $E_2^2 = e_4 + e_3$, $E_3^2 = e_4 + e_6$, $E_4^2 = e_4 + e_7$, $E_5^2 = e_4 + e_8$, $E_6^2 = e_4 + e_9$, $E_7^2 = e_4 + e_{10}$ and $E_8^2 = e_4 + e_{11}$ be another set of eight digits. The eight parity-check sums orthogonal on E_1^2, $E_2^2, \dots$, E_8^2 are

$\{1+14$, $13+22$, $0+3+12$, $5+11+16$, $8+9+17$, $7+21, 6+$
$8+23$, $10+15+19+20\}$, $\{1+2+17$, $13+20$, $5+11+23$, $0+10+$
22, $9+21$, $8+16+19$, 7, $6+14+15+18\}$, $\{5+23$, $0+18+$
22, $2+16+19$, $15+20$, $1+12+21$, $10+13+17$, $9+14$, $6+7$
$+11\}$, $\{3+5+19$, $0+16+19$, $14+18+20$, $11+12+17$, $2+$
21, $1+8+9$, $6+13+15\}$, $\{3+18+23$, $0+17$, $5+16$, $15+22$,
$1+12$, $9+14+21$, $7+10+19$, $2+6+13+20\}$, $\{1+16+19$, 2
$+13+23$, $3+12$, $11+17$, $10+21+22$, $5+8+20$, $7+15+18$, 6
$+14\}$, $\{2+3+23$, $0+5+17$, 16, $14+15$, $12+13$, $1+11+22$,
$9+20$, $7+8+19\}$, and $\{0+5+20$, $19+23$, $16+18$, $2+15+$
22, $10+13+14$, $9+17$, $6+7$, $1+3+12+21\}$,

respectively. From these orthogonal sums, E_1^2, E_2^2, ..., E_8^2 can be correctly estimated provided that there are no more than four errors in the error vector. We see that E_1^2, E_2^2, ..., E_8^2 are orthogonal on e_4. Therefore, e_4, and thereafter e_2, e_3, e_6, e_7, e_8, e_9, e_{10}, and e_{11}, can be estimated from these sums.

Thus, using the above parity sum equations, e_0, e_1, ..., e_5 can be estimated correctly if no more than four errors are present in the received vector. Hence, the (24,6,9) code is a two-step sufficiently orthogonalizable code.

5.8   Appendix B

Weight Distribution of Some Codes Obtained from the Aperiodic Convolution Algorithm

Code	Weight											
Parameters	0	7	8	9	10	11	12	13	14	15	16	17	18
(18,5,7)	1	5	6	6	6	5	3						
(20,6,7)	1	5	8	13	11	7	11	7	1				
(22,5,9)	1			4	7	8	3	4	5				
(24,6,9)	1			4	8	16	12	4	8	8	3		
(24,4,11)	1					2	6	4	2	1			
(26,7,9)	1			5	13	21	17	15	15	15	18	8	
(26,5,11)	1					2	6	12	4	2	5		
(28,6,11)	1					4	8	16	12	4	7	8	4

Note: All blank entries are zeros.

Chapter 6

A New Error Control Scheme

This chapter introduces the concept of a generalized hybrid Automatic-Repeat-Request (ARQ) scheme for adaptive error control in digital communication systems. This scheme utilizes the redundant information available upon successive retransmissions in an efficient manner so as to provide high throughput during poor channel conditions. Either class of linear codes already discussed may be used for the generalized hybrid ARQ (GH-ARQ) system application; however, we will concentrate here on those linear codes resulting from a bilinear algorithm for *aperiodic* convolution. The main feature of this class of codes (and those derived from a bilinear algorithm for periodic convolution) is that the encoder/decoder configuration does not change as the length of the code is varied. As a result, the receiver uses the *same decoder* for decoding the received information after every retransmission while the error-correcting capability of the code increases, thereby leading to both improved performance and minimum complexity for the overall system implementation. It is hoped that this chapter will illustrate just one way in which we may find practical reward from the efforts of past chapters.

6.1 ARQ Schemes

There are two fundamental techniques for error control in digital communication systems: Forward-Error-Control (FEC) and Automatic-

Repeat-Request (ARQ) schemes [6.1]. In a digital communication system using FEC, the transmitter employs an error-correcting code to add redundant digits to the information digits in a manner so as to correct the error patterns that are most likely to occur during transmission. At the receiving end, the decoder attempts to recover the information digits from the received digits in a way so as to maximize a certain system performance measure.

In digital communication systems using ARQ, the data to be transmitted is first organized in blocks after which an error-detecting code is used to encode each block in order to achieve the required error detection capability. No error correction is performed by the decoder. When an error is detected in a block, a request for retransmission is made through a reverse (or feedback) channel. The transmitter is informed of whether a block is correctly or incorrectly received by an acknowledgement (ACK) or nonacknowledgement (NACK), respectively. If an ACK is received at the transmitter, a new block is transmitted; whereas, if a NACK is received, the same block is transmitted again. With this procedure, a block is delivered to the user only when no errors are detected by the receiver.

Both of these error-control schemes, FEC and ARQ, have their relative advantages and disadvantages. One important measure of performance of such systems is their *throughput efficiency* [6.2]. The throughput efficiency is defined as the ratio of the number of information bits to the total number of bits transmitted. Since there are no retransmissions in a FEC system, the throughput efficiency is constant and set by the code rate, regardless of the channel transmission conditions. The most severe drawback of an ARQ system is its throughput efficiency. The throughput efficiency of an ARQ system depends strongly on the number of requested retransmissions, that is, on channel quality and, therefore, falls rapidly with increasing channel error rate [6.3, 6.4].

Another important measure of performance is the *system reliability*. In FEC systems, the reliability of the received data is very sensitive to any degradation in the channel conditions and the selection of an appropriate error-correcting code depends on the detailed knowledge of the error statistics of the channel [6.1]. If an appropriate choice is made for the error-detecting code, the probability of an undetected error can be made very small [6.5]. As the probability of a decoding error for an error-correcting code is much greater than the probability of an undetected error for an error-detecting code, ARQ systems are far more reliable than FEC systems. Furthermore, a code used for error detection is not very sensitive to the actual error patterns and it can detect the vast majority of the error patterns. Consequently, unlike FEC systems, the use of ARQ error control is effective on most channels.

The cost of an ARQ system is substantially lower than that of a FEC system. This is due to the fact that error detection is, by its nature, a much simpler task than error correction. Also, error detection with retransmission is *adaptive*, i.e., transmission of redundant information is increased when errors occur. This makes it possible under certain circumstances to achieve better performance with an ARQ system than is theoretically possible with a FEC system.

From the above discussion, it is clear that the ARQ scheme can provide high system reliability reasonably independently of the channel quality. As the channel becomes noisier, however, the requests for retransmission increase, leading to a reduced throughput. On the other hand, FEC techniques provide a constant throughput regardless of the channel quality, but the system reliability falls as the channel degrades.

In situations where the channel error rate is too high to guarantee desired throughput using ARQ and where the required system reliability is too high to be achieved by FEC alone, a combination of FEC and ARQ systems may be attractive. Such a scheme is termed hybrid ARQ [6.5, 6.6]. In principle, the hybrid scheme combines the advantages of both techniques. A combination of correction of the most frequent error patterns (FEC) and detection coupled with retransmission for less frequent error patterns (ARQ) has the potential to provide high throughput efficiency as well as high system reliability.

In this chapter, we employ the new class of linear error-correcting codes introduced in Chapter 2 and discussed in Chapter 5 for the hybrid ARQ application. These codes are characterized by their modular structure which, in turn, can be used to design codes with variable minimum distance having the same encoding/decoding procedures. The interested reader may consult [6.7-6.9] for more details.

6.2 Hybrid ARQ Schemes

In this section, the various hybrid ARQ schemes that can be used for error control are presented, and their main features are highlighted.

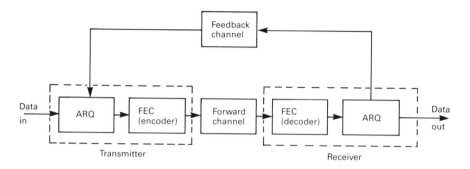

Figure 6.1: Type-I Hybrid ARQ System.

Depending on the retransmission protocols, the hybrid ARQ schemes can be based upon any one of the three basic ARQ schemes, i.e., stop-and-wait ARQ, go-back N or continuous ARQ, and selective-repeat ARQ [6.5].

There are a number of factors that require analysis in order to select an appropriate scheme for the error control problem. The main factors are the channel characteristics and the cost and complexity of the system required to implement the scheme.

As stated in Section 6.1, hybrid ARQ schemes can be used to overcome the drawbacks of FEC and ARQ systems. Such hybrid schemes can be broadly classified as [6.3]:

1. Type-I Hybrid ARQ Schemes and

2. Type-II Hybrid ARQ Schemes.

Type-I Hybrid ARQ Schemes. A type-I hybrid ARQ system employs a code for error correction (FEC) and error detection (ARQ) as shown in Figure 6.1. At the receiver, the decoder first attempts to correct the received block, after which error detection is carried out on the block. If an uncorrectable error pattern is detected, the received block is rejected and a NACK is sent to the transmitter. The transmitter, upon receiving a NACK, retransmits the same block. This procedure is continued until the block is successfully accepted by the receiver. In this technique, an erroneous block is delivered to the user only if error patterns in the decoded block cannot be detected by the error-detecting code.

Since a type-I ARQ scheme uses codes for error correction as well as error detection, it requires more parity-check digits than a code used for error detection only, as in the ARQ schemes. Consequently, for good channels the throughput of a type-I hybrid ARQ scheme may be lower than that of the corresponding ARQ scheme. When the channel error rate is high, however, a type-I hybrid ARQ scheme provides higher throughput than the corresponding ARQ scheme, because its error-correcting code is designed to correct the most frequently occurring channel error patterns. For this reason, type-I hybrid ARQ schemes are best suited for communication over channels whose characteristics are known *a priori* to be fairly constant.

For communication systems in which the channel characteristics change, a type-I hybrid ARQ system may not be very efficient. When the channel is quiet (that is, errors are introduced very infrequently), no error correction is required and, therefore, the extra parity-check digits for error correction included in each block are not needed. When the channel is noisy, i.e., the errors are introduced frequently, the error-correcting code may become inadequate. This, in turn, increases re-transmissions, lowering the throughput. It can alternately be argued that it is difficult in practice to design good error-correcting codes for channels having varying characteristics. An example of such a channel is the satellite channel which is very quiet in good weather and behaves poorly during rain.

Type-II Hybrid ARQ Schemes. For a channel having *nonstationary* behavior, it might be desirable to utilize an adaptive hybrid ARQ scheme. Error detection with retransmission is *adaptive,* i.e., transmission of redundant information is increased when errors occur. This forms the basis of type-II hybrid ARQ schemes. The concept of variable redundancy codes was first introduced by Mandelbaum [6.11], and based on this concept, a type-II hybrid ARQ scheme was proposed by Metzner [6.12], Lin et al [6.3], and Wang et al [6.13].

In the type-II hybrid ARQ system, the parity-check digits for error correction are sent to the receiver only when they are needed. The type-II hybrid scheme uses two linear codes denoted by C_0 and C_1, where C_0 is an (n, k) error-detecting code and C_1 is a $(2n, n)$ half-rate *invertible* code, designed for error correction only. A $(2n, n)$ code is said to be an invertible code if, knowing only the n parity-check digits of a codeword, the associated n information digits can be *uniquely* determined by an inversion process.

Let I denote the block of length n formed on the basis of the information vector of length k and the (n, k) code C_0. Corresponding to the block I, a parity block $P(I)$, also of length n, is formed using I and the code C_1. Now, I is transmitted; the corresponding received block is denoted by $\tilde{I}$. When $\tilde{I}$ is received, the receiver attempts to detect the presence of errors using C_0. If no errors are detected, $\tilde{I}$ is assumed error-free and an ACK is sent to the transmitter. If the presence of errors is detected by C_0, the received block is stored in a buffer and a NACK is sent.

Upon receiving a NACK, the transmitter sends the parity block, $P(I)$; the corresponding received block is denoted by $\tilde{P}(\mathrm{I})$. When $\tilde{P}(\mathrm{I})$ is received, the receiver takes the inverse of $\tilde{P}(\mathrm{I})$, denoted by $\mathrm{I}(\tilde{P})$, based on C_1. If $\tilde{P}(\mathrm{I})$ is error-free, $\mathrm{I}(\tilde{P})$ is the same as the original block I and is a codeword in C_0. Therefore, once $I(\tilde{P})$ is calculated, the receiver attempts to detect errors in $I(\tilde{P})$ based on C_0. If no errors are detected, $I(\tilde{P})$ is assumed error-free and an ACK is sent. If errors are detected in $I(\tilde{P})$, then the blocks $\tilde{P}(I)$ and $\tilde{I}$ are used *together* for error correction based on the $(2n, n)$ code C_1. Let I^* denote the decoded block after the error correction process. Now, I^* is checked based on C_0. If no errors are detected, then I^* is assumed error- free and an ACK is sent. If errors are detected in I^*, the receiver discards the block $\tilde{I}$, stores the block $\tilde{P}(I)$, and sends a NACK. The second retransmission is the block I itself. The error detection/correction procedure is then repeated. The retransmissions continue until the block is successfully received. The retransmissions are alternate repetitions of the block I and the parity block $P(I)$.

Depending on the choice of the error-correcting code C_1, two variations of the above scheme are available. Most approaches use codes over GF(2). In [6.12], Metzner proposes a scheme wherein small sub-blocks of the original n-bit block are taken to be the data bits of a rate one-half code. The choices for the code are stated to be limited to the $(6,3,3)$, $(8,4,4)$ Reed-Muller, and $(16,8,5)$ codes. In [6.13], Wang et al describes a scheme wherein the code C_1 operates on the complete data block and, consequently, it is a code having large length, dimension, and distance properties. It is worthwhile to mention here that for ARQ applications, a large value $(n \sim 500)$ is usually selected for n [6.3].

It is clear from the above description that the overall performance

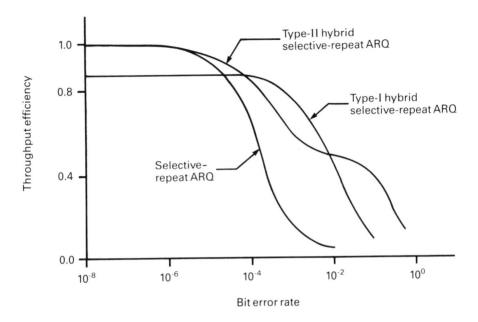

Figure 6.2: Throughput Efficiency of Various ARQ Schemes [6.2].

of the type-II hybrid ARQ system depends very strongly on the choice of the error-correcting code C_1. In this respect, the scheme performs better for the codes given in [6.13], as compared to the subblock encoding approach of [6.12]. The complexity of decoding a large $(2n, n)$ code is, however, much greater than the complexity of decoding a small code. In addition, long codes must decode after hard digit decisions, whereas soft-decision decoding of block codes can be used to make a combined hard decision for each of the subblocks in the subblock encoding approach. This interesting point will be pursued in Section 6.8. Furthermore, since long codes generally have large decoding delays, it may not be feasible to use such a code in high data rate communication systems.

In Figure 6.2, typical plots of throughput efficiency versus channel bit error rate for the selective-repeat ARQ, type-I, and type-II hybrid selective-repeat ARQ schemes are presented for comparison [6.2]. As can be observed from the figure, the throughput of the selective-repeat ARQ and type-I hybrid ARQ technique decreases rapidly as the channel begins to degrade, while the throughput of the type-II hybrid ARQ technique falls less rapidly as it approaches 0.5. This is due to the fact that error correction is performed upon the first retransmission in type-II hybrid ARQ systems and, therefore, the probability of further retransmissions is considerably reduced.

6.3   Generalized Type-II Hybrid ARQ Schemes

In this section, we generalize the type-II hybrid ARQ scheme described in the previous section. It has been stated earlier that error detection with retransmission is adaptive. The first retransmission of a block provides the receiver with n redundant digits, and this forms the basis of the type-II hybrid ARQ scheme of the previous section. A natural question arises at this state: "What is the optimal way of using the redundant information when more than one retransmission is requested?" For example, in the type-II ARQ scheme, the second retransmission is the same as the information block. Clearly, the performance of the system can be further improved if the second retransmission is another parity block which may be used to form a $(3n, n)$ error-corrrecting code. Such a scheme can be generalized to any number of retransmissions before the transmitter resends the blocks in repetition once again. We will refer to this scheme as a generalized type-II hybrid ARQ (GH-ARQ) scheme [6.7-6.9].

The GH-ARQ scheme also uses two codes; the first is a high rate (n, k) code C_0 which is designed for error detection only and the second is the code C_1, used adaptively for error correction. The code C_1 is an (mn, n) error-correcting code having distance d, selected in such a way that its generator matrix can be partitioned into m subblocks each of dimension $(n \times n)$. The integer m will be referred to as the *depth* of the code. Let the generator matrix of C_1 be denoted by G and the subblocks of G be denoted by $G_1, G_2, \ldots, G_m$. Then, G can be written as

$$G = [G_1 \mid G_2 \mid \cdots \mid G_m]. \tag{6.1}$$

For the code C_1 to be useful for the application under consideration, it is assumed that the subcode $C_1^{(i)}$ with the generator matrix $G^{(i)}$, where $G^{(i)} = [G_1 \mid G_2 \mid \cdots \mid G_i]$, has minimum distance $d_i < d_j$ for all $1 \leq i < j \leq m$. The depth of the subcode $C_1^{(i)}$ is i by definition. Note that the code $C_1^{(m)}$ is the same as the code C_1.

Such a code can be incorporated into a GH-ARQ scheme as follows. Let I denote a block which is formed based on the information vector and the (n, k) code C_0. The mn-digit long codeword is formed using I and the (mn, n) code C_1. Let such a codeword be represented by $\mathbf{c}^T = [\mathbf{c}_1^T | \mathbf{c}_2^T | \cdots | \mathbf{c}_m^T]$, where the vector $\mathbf{c}_i$ corresponds to $G_i, i = 1, 2, \ldots, m$, and has length n. We know that the data block I can be uniquely recovered from knowledge of $\mathbf{c}_i$ *iff* the corresponding $(n \times n)$ matrix G_i is invertible. Therefore, the generator matrix G_1 is assumed to be invertible so that the data block can be computed from $\mathbf{c}_1$ alone (first transmission). In addition, it is desirable (although, not necessary) that the $(n \times n)$ matrices $G_i, i = 2, \ldots, m$, be invertible as well. This is particularly important for the case in which a burst of errors might destroy one of the transmissions yet leave the other transmissions relatively error-free. Codes having these desired properties can be obtained from either the periodic or aperiodic bilinear algorithms associated with the bilinear form introduced in Section 2.4.

For the block I to be transmitted, the sequence of blocks which are transmitted to the receiver until the block is successfully accepted is given by $\mathbf{c}_1, \mathbf{c}_2, \ldots, \mathbf{c}_m, \mathbf{c}_1, \mathbf{c}_2, \ldots, \mathbf{c}_m, \mathbf{c}_1, \ldots$. Upon receiving a block, say $\mathbf{c}_i$, the receiver adopts one of the following two strategies:

1. Invert $\mathbf{c}_i$ (if the associated G_i is invertible); perform error detection. If required, decode using $\mathbf{c}_i$ and the previously received blocks; perform error detection again, or

2. Decode using $\mathbf{c}_i$ and the previously received blocks; perform error detection.

An ACK is sent to the transmitter whenever no errors are detected. For the above transmitted sequence of blocks, the receiver performs error correction on the basis of the code sequence, $C_1^{(2)}, C_1^{(3)}, \ldots, C_1$, $C_1, \ldots$ having the minimum distance sequence $d_2, d_3, \ldots, d, d, \ldots$, respectively. Thus, with each retransmission, a code having a larger minimum distance, that is, a larger error-correcting capability, is used for error correction until the code C_1 is reached.

Note that the type-II hybrid ARQ scheme described in Section 6.2 is a special case of the GH-ARQ method presented here, for which the depth $m = 2$, the matrices G_1 and G_2 are invertible, and, furthermore, the matrix G_1 is an identity matrix.

Complexity of the GH-ARQ Scheme. We briefly analyze the complexity of the GH-ARQ scheme described above. To begin with, for any block I to be transmitted, the transmitter computes a codeword of length mn based on C_1. A buffer of length mn bits is required at the transmitter and the receiver for each of the blocks to be transmitted. The transmitter also requires an encoder only for the code C_1. The receiver requires a decoder for *each* of the codes $C_1^{(2)}, \ldots, C_1^{(m-1)}, C_1$. Also, an inverter circuit is needed for each of the $\mathbf{c}_i, i = 1, 2, \ldots, m$.

From this rudimentary discussion of the system complexity, the complexity of implementing the GH-ARQ scheme may appear to be prohibitive. Specifically, the configuration of the receiver seems very complex as it requires potentially complex decoders for *more* than one code, namely $C_1^{(2)}, \ldots, C_1^{(m-1)}, C_1$, and, if the decoding procedure for these codes is not the same, the cost of having $(m-1)$ decoder circuits at the receiver may offset any potential gain in system performance. Alternatively, if the decoder configuration of the codes is the *same*, then the GH-ARQ scheme may offer a significant advantage over the type-II ARQ systems, while keeping the additional system complexity as low as possible. One class of linear codes having the necessary properties has previously been discussed in Chapter 5 and will henceforth be referred to as the class of KM codes. In the next section, we present a salient review of the important features of the KM codes. These codes are derived from bilinear algorithms for aperiodic convolution and are associated with the duality of the problem for periodic convolution.

6.4   A Brief Review of the KM Codes

The KM codes are obtained from algorithms for computation of aperiodic convolution of two indeterminate sequences of lengths k and d over $GF(p)$, as described in Section 2.4 and Chapter 5. The parameter k is the dimension of the code generated, and d is the design minimum distance of the code. The length of the code n is equal to the multiplicative complexity of the algorithm used for the computation. Our principle interest here will be with design over the field of constants $GF(2)$.

Consider the aperiodic convolution of two indeterminate sequences, $z_i, i = 0, 1, \ldots, k-1$, of length k and $y_j, j = 0, 1, \ldots, d-1$, of length d over $GF(2)$. Let the resulting sequence be $\phi_t, t = 0, 1, \ldots, N-1$, where $N = k + d - 1$. We have shown that if A^T $(C\mathbf{z}^T \bullet B\mathbf{y})$ is the algorithm over $GF(2)$ for computation of the aperiodic convolution, then the matrix C represents the generator matrix of an $(n, k, \bar{d})$ binary linear code, where $\bar{d} \geq d$. The matrices A^T, C^T, and B have dimensions $(N \times n), (n \times k)$, and $(n \times d)$, respectively, and $\bullet$ represents component-by-component multiplication of vectors. The column vectors $\mathbf{z}$ and $\mathbf{y}$ are defined as the indeterminate vectors $[z_0 z_1 \cdots z_{k-1}]^T$ and $[y_0 y_1 \cdots y_{d-1}]^T$, respectively.

Define the generating polynomial of a sequence $x_r, r = 0, 1, \ldots, \ell-1$, as

$$X(u) = \sum_{r=0}^{\ell-1} x_r u^r. \tag{6.2}$$

Let $Z(u)$, $Y(u)$, and $\Phi(u)$ denote the generating polynomials of the sequences $z_i, i = 0, 1, \ldots, k - 1; y_j, j = 0, 1, \ldots, d - 1;$ and $\phi_t, t = 0, 1, \ldots, N - 1$, respectively. The aperiodic convolution can alternatively be computed as the polynomial product

$$\Phi(u) = Z(u)\, Y(u). \tag{6.3}$$

Since $\Phi(u)$ is of degree $N - 1 = k + d - 2$, $\Phi(u)$ is unchanged if it is defined modulo any polynomial $P(u)$ of degree at least N. That is,

$$\Phi(u) = Z(u)\, Y(u) \bmod P(u), \ \deg [P(u)] \geq N.$$

If $P(u)$ is the product of L relatively prime polynomials $P_i(u), i = 1, 2, \ldots, L$, that is, $P(u) = \prod_{i=1}^{L} P_i(u)$, $\Phi(u)$ can be computed by first reducing the polynomials $Z(u)$ and $Y(u)$ modulo $P_i(u)$, where

$$\begin{aligned} Z_i(u) &\equiv Z(u) \bmod P_i(u), \\ Y_i(u) &\equiv Y(u) \bmod P_i(u), \quad i = 1, 2, \ldots, L. \end{aligned} \tag{6.4}$$

The polynomial $\Phi(u)$ is then obtained by computing the L polynomial products $Z_i(u)Y_i(u)\bmod P_i(u)$ and using the CRT to uniquely reconstruct $\Phi(u)$ from the products modulo $P(u)$. Since the multiplicative complexity of the aperiodic convolution also determines the length n of the code, a worthwhile objective is to develop efficient algorithms with as low a multiplicative complexity as possible for given values of k and d.

Let $M(P(u), N)$ denote the multiplicative complexity of the polynomial product $Z(u)Y(u) \bmod P(u)$, where $\deg[P(u)] = N$. The multiplicative complexity of the algorithm can often be further reduced by suitably modifying the above described basic procedure to permit *intentional* wraparound of the polynomial coefficients, as discussed in Section 3.2.3. For example, if $M(P'(u), N-1) < M(P(u), N) - 1$, where $\deg[P'(u)] = N - 1$, then it is more efficient to calculate $\Phi(u)$ from the polynomial product $Z(u)Y(u) \bmod P'(u)$ and one extra multiplication, i.e., $z_{k-1} y_{d-1}$. This procedure is easily generalized to more than one wraparound.

Let the number of polynomial coefficients which are computed using the wraparound be s, and let M_s be the number of extra multiplications required in such a procedure. An optimal choice for the degree D of the polynomial $P(u)$ and the integer s is one such that $N = D + s$ and $M(P(u), D) + M_s$ has the smallest value possible. Also, $M(P(u), D)$ depends on the form of the polynomial $P(u)$. As seen in Chapter 5, the generator matrix of the code is block structured with $L + 1$ blocks, the first L blocks corresponding to the L mutually prime factor polynomials $P_i(u)$ and the last block corresponding to the wraparound computation. We illustrate the procedure for linear code generation by the following example.

EXAMPLE 1.   Consider the $(12,4,5)$ code obtained from an aperiodic convolution algorithm over $GF(2)$. Since $N = k + d - 1 = 8$, we have $k + d = 9$. The generating polynomials are $Z(u) = z_0 + z_1 u + \cdots + z_3 u^3$ and $Y(u) = y_0 + y_1 u + \cdots + y_4 u^4$. We choose $P(u) = u^2(u^2 + 1)(u^2 + u + 1)$ and $s = 2$, to compute the aperiodic convolution $\Phi(u) = Z(u)Y(u)$ (Appendix C of Chapter 3 is helpful in this regard). Let $P_1(u) = u^2, P_2(u) = u^2 + 1$, and $P_3(u) = u^2 + u + 1$. Each of the computations $Z_i(u)Y_i(u) \bmod P_i(u), i = 1, 2, 3$, and the wraparound require three multiplications. The generator matrix of the corresponding $(12,4,5)$ code is given by

$$M = \begin{bmatrix} 1 & 0 & 1 & 1 & 0 & 1 & 1 & 1 & 0 & 0 & 0 & 0 \\ 0 & 1 & 1 & 0 & 1 & 1 & 0 & 1 & 1 & 0 & 0 & 0 \\ 0 & 0 & 0 & 1 & 0 & 1 & 1 & 0 & 1 & 0 & 1 & 1 \\ 0 & 0 & 0 & 0 & 1 & 1 & 1 & 1 & 0 & 1 & 0 & 1 \\ & \text{for} & & & \text{for} & & & \text{for} & & & \text{wrap-} & \\ & P_1(u) & & & P_2(u) & & & P_3(u) & & & \text{around} & \end{bmatrix}.$$

$$(6.5)$$

It is worthwhile to once again mention that the design minimum distance of the above code is 5, while the actual minimum distance is 6.

Decoding Algorithm. The decoding algorithm for KM codes utilizes the block structure of the codes as discussed in Chapter 5. The computation for each of the factor polynomials can be performed in a way that the corresponding partition of the codeword is also a codeword of an $(n_i, \sigma_i, \sigma_i)$ code, where n_i is the length of the partition and $\deg[P_i(u)] = \sigma_i$. The decoder, therefore, uses decoders for each of the partitions as a basis for the decoding of the overall code. Note that the partitions are relatively small compared to the code itself.

Once decoding for each partition is complete, an estimate of the residue polynomial associated with each of the partitions is obtained. Since the information polynomial is of degree $(k-1)$, knowledge of the residue polynomials whose sum of degrees is k is sufficient to *uniquely* reconstruct the information polynomial using the CRT. As the received vector may contain transmission errors, the decoder forms estimates of the information polynomial from all the possible combinations of the residue polynomials using the CRT and accepts the estimate for which the associated codeword differs from the received vector in the least number of digits as a valid information polynomial. A block diagram of the general decoder configuration of the KM codes is shown in Figure 5.6 of Chapter 5.

If we discard those columns of the generator matrix of a KM code corresponding to the computation modulo a polynomial $P_i(u)$, then the resulting matrix is the generator matrix of an $(n - n_i, k, d - \sigma_i)$ KM code. The decoder of such a code can be obtained by disabling (1) the decoder for the partition corresponding to the deleted columns and (2) the reconstruction which utilizes the residue modulo $P_i(u)$. A suitable adjustment of the threshold in the selector is also required.

We conclude that for a given value of k, it is possible to decrease (increase) the minimum distance d of the code by simply deleting (adding) columns of (to) the generator matrix, thereby leading to an appropriate decrease (increase) in the length n of the code. In the latter case, however, care must be taken in appropriately selecting the columns to be added; this point is discussed in Section 5.4.2. The encoder/decoder design for all such codes remains essentially unaltered; therefore, the system designer can incorporate a large number of such codes in a single encoder/decoder configuration.

6.5   A GH-ARQ Scheme Based on KM Codes

We now describe a GH-ARQ scheme that employs KM codes for error correction. To be specific, we take the example of one particular code and illustrate the complete scheme for such a code. Such a scheme, however, is general in nature, and in Appendix A, we present the generator matrices of several KM codes along with their useful features for such an application.

The generator matrix of the (12,4,6) KM code is given in (6.5). Rearranging the columns of the generator matrix M, we obtain

$$
M = \begin{bmatrix}
1 & 1 & 0 & 1 & 1 & 1 & 0 & 0 & 0 & 1 & 0 & 0 \\
0 & 0 & 1 & 0 & 1 & 1 & 1 & 0 & 1 & 1 & 0 & 0 \\
0 & 1 & 0 & 1 & 1 & 0 & 1 & 0 & 0 & 0 & 1 & 1 \\
0 & 0 & 1 & 1 & 1 & 1 & 0 & 1 & 0 & 0 & 0 & 1
\end{bmatrix} = [M_1|M_2|M_3]. \quad (6.6)
$$

It is easy to verify that the matrices M_1, M_2, and M_3 are *invertible*. Also, the matrix $[M_1|M_2]$ is the generator matrix of the (8,4,3) KM code, since it corresponds to the polynomial product $Z(u)Y(u)$, where $Z(u) = z_0 + z_1 u + \cdots + z_3 u^3$, $Y(u) = y_0 + y_1 u + y_2 u^2$, $P(u) = u(u^2 + 1)(u^2 + u + 1)$, and $s = 1$.

Let us assume that we wish to use the above code in a GH-ARQ system. Choosing an (n, k) code C_0 for error detection, we proceed as follows. Define three matrices G_1, G_2, and G_3 each of dimension $(n \times n)$ as

$$
G_i = M_i \otimes I_{n/4}, \quad i = 1, 2, 3, \quad (6.7)
$$

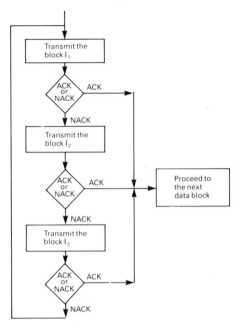

Figure 6.3: Transmission Procedure for a GH-ARQ Scheme Using Depth 3 Codes for Error Correction.

where $I_{n/4}$ is an identity matrix of order $n/4$ and $\otimes$ represents the Kronecker product of two matrices. It is worthwhile to note here that since the M_i, $i = 1, 2, 3$, are invertible, the matrices G_i, $i = 1, 2, 3$, are also invertible and G_i^{-1} is given by

$$G_i^{-1} = M_i^{-1} \otimes I_{n/4}, \quad i = 1, 2, 3. \tag{6.8}$$

Let ℓ denote the vector corresponding to the block I to be transmitted. The transmitter computes three $(n \times n)$ vectors

$$c_i^T = \ell^T G_i, \quad i = 1, 2, 3. \tag{6.9}$$

Since the $G_i, i = 1, 2, 3$, are invertible, the data block I can be uniquely computed from the $c_i, i = 1, 2, 3$. Let I_1, I_2, and I_3 denote the blocks associated with the vectors c_1, c_2, and c_3, respectively. The transmission procedure for the block I is shown in Figure 6.3.

Let R_i denote the received block corresponding to the block I_i and let $\hat{I}_i$ be an estimate of the block I based on R_i and the matrix $G_i, i =$

$1, 2, 3$. If no errors have taken place during transmission, then $R_i = I_i$ and, consequently, $\hat{I}_i = I$, $i = 1, 2, 3$. The receiver configuration for the block I is given in Figure 6.4. In Figure 6.4, the integers i and j correspond to the total number of transmissions for the block I and the most recently received block, respectively.

It was implicitly assumed above that the length of the block I was an integral multiple of 4, the dimension of the KM code used. In general, the length of a block in an ARQ scheme may be restricted due to other system constraints.

6.6 GH-ARQ Error Detection

In this section, we analyze the error detection capability of the GH-ARQ technique. This is an important parameter for the performance of the system. Let P_e denote the probability of an undetected error for the (n, k) error detection code C_0. If C_0 is properly chosen, P_e is upperbounded by

$$P_e < [1 - (1 - \varepsilon)^k]2^{-(n-k)}, \quad 0 < \varepsilon < \frac{1}{2}, \tag{6.10}$$

where ε is the bit error rate of the channel [6.13]. If a code that satisfies the above bound is used for ARQ systems, then P_e can be made very small by using an appropriate number of parity-check digits. The bound is an *existence* bound, and very few codes have been shown to satisfy it theoretically [6.14]. It is intuitively seen, however, that there is an abundance of codes that will satisfy the bound. We note that the reliability of an ARQ scheme depends strongly on the existence of codes for which P_e can be made arbitrarily small. In the following, we examine the probability of undetected error for each transmission in a GH-ARQ system as compared to the probability of undetected error for each transmission in an ARQ scheme using the code C_0.

Let the column vector $\mathbf{r}$ denote the received vector and let the $[(n - k) \times n]$-dimensional matrix H be the parity-check matrix of the code C_0. The procedure for error detection corresponds to computing the vector-matrix product $\mathbf{r}^T H^T$. If the product is zero, the vector $\mathbf{r}$ is a codeword in C_0 and is declared error-free. In the GH-ARQ system, the receiver first takes the inverse of the received vector $\mathbf{r}$ in order to obtain an estimate of the codeword I and then examines it using C_0.

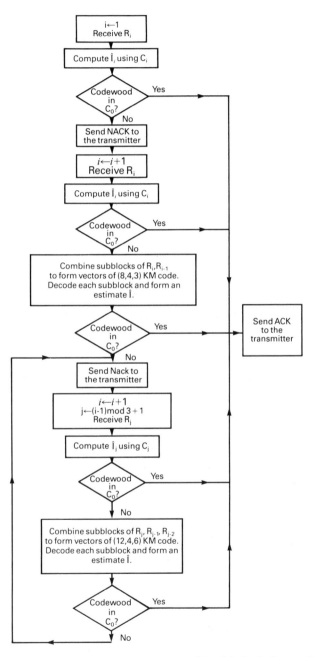

Figure 6.4: Receiver Operation for a GH-ARQ Scheme Using the (12,4,6) KM Code for Error Correction.

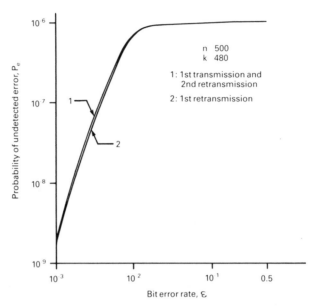

Figure 6.5: Probability of Undetected Error for the GH-ARQ Scheme Using the (12,4,6) KM Code for Error Correction.

Such a procedure can be mathematically described as the vector-matrix product $\mathbf{r}^T G_i^{-1} H^T$, where the subscript i represents the ith transmission. Let $\tilde{H}_i^T = G_i^{-1} H^T$; we can, equivalently, state that, in a GH-ARQ system, an error pattern $\mathbf{e}$ in the ith transmission of a block will be undetectable *iff* $\mathbf{e}$ is a codeword in the linear code having $\tilde{H}_i$ as its parity-check matrix. We now present two examples which satisfy the bound (6.10).

EXAMPLE 2. Let C_0 be the (500,480) code obtained by shortening the (1023, 1003, 5) BCH code and let C_1 be the (12,4,6) code described in Section 6.5. Figure 6.5 shows the probability of undetected error when C_0 is used for error detection in a GH-ARQ scheme that uses the (12,4,6) code for error correction. Note that in this case there are three plots for the probability of undetected error; these correspond to the first transmission and first and second retransmissions.

EXAMPLE 3. For the same code C_0 as in Example 2, Figure 6.6 shows the probability of undetected error for successive transmissions that use the (15,5,5) code of Appendix A for adaptive error correction.

We observe that in both cases, the probability of undetected error is a monotonic function of $\varepsilon, 0 < \varepsilon < \frac{1}{2}$, a condition necessary and sufficient for a code to satisfy the bound of (6.10) [6.14].

Burst Error Detection Capability of the GH-ARQ Scheme. For ARQ applications, the error detection code C_0 is usually chosen as an (n, k) cyclic code due to its burst error detection capability. Every (n, k) cyclic code can detect any burst of length $(n-k)$ or less, as no codeword of an (n, k) cyclic code is a burst of length $(n - k)$ or less [6.1].

A natural question to consider is the burst error detection capability of the codes used in the GH-ARQ scheme. For the GH-ARQ scheme, the receiver multiplies subblocks of the received vector with the matrix inverse M_i^{-1}. If there are errors in a subblock, these errors are, therefore, contained within the subblock itself after the inversion. For the worst case, we assume that if there is an error in a subblock, *all* the digits in the subblock are in error after inversion. If l' is the length of each subblock and n' is the number of subblocks that are affected by a burst of errors, then, after inversion, the length of the burst is at most $l'n'$. For a burst to be detectable, we have $l'n' \leq (n - k)$ or $n' \leq \lfloor (n - k)/l' \rfloor$. The maximum value of n' is, therefore, given by $n'_{max} = \lfloor (n - k)/l' \rfloor$.

It is a simple exercise to establish that the maximum length of a burst that affects only n'_{max} number of subblocks is given by $l'(n'_{max} - 1)+1$. The burst error-detecting capability is, therefore, underbounded by $l'(n'_{max} - 1) + 1$. For example, if $(n - k) = 20$ and $l' = 4$ (as is the case for the (12,4,6) code), then $n'_{max} = 5$ and the burst error-detecting capability is at least 17.

6.7 Reliability and Throughput

In this section, we analyze the reliability and throughput of the GH-ARQ scheme. These are two important parameters associated with the performance of any ARQ scheme.

Reliability. It has been established that the type-II hybrid ARQ system provides the same order of reliability as an ARQ system [6.13].

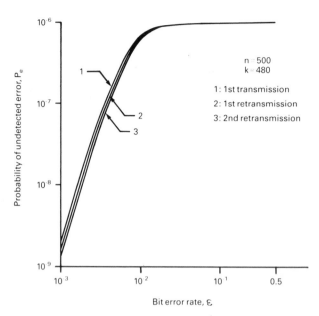

Figure 6.6: Probability of Undetected Error for the GH-ARQ Scheme Using the (15,5,5) KM Code for Error Correction.

If the code C_o is chosen in such a way that the probability of undetected error in any transmission of the GH-ARQ system is the same as the probability of undetected error in any transmission of an ARQ system, then using similar arguments, we can show that the GH-ARQ system also provides the same order of reliability as an ARQ system. A sufficient condition for this is that the probability of undetected error specified by the parity-check matrix $\tilde{H}_i^T = G_i^{-1} H^T$ satisfies (6.10). The expressions of Appendix B may be employed to assess GH-ARQ system reliability.

Throughput. Since selective-repeat ARQ is the most efficient ARQ scheme, we consider the throughput of the GH-ARQ scheme in the selective-repeat mode. The throughput of such a scheme depends on the buffer size in the receiver, and in this regard, we restrict our attention to the infinite buffer case. Also, it is assumed that the feedback channel is noiseless.

Let A_o^c, A_o^d, and A_o^e be the events that a data block contains no errors, detectable errors, and undetectable errors, respectively, in its first transmission. Upon the ith retransmission, let E_i^c be the event that the receiver recovers the data block correctly; let E_i^d be the event that the receiver cannot recover the data block, detects the presence of errors, and requests the next retransmission; and let E_i^e be the event that the receiver accepts an erroneous data block. If V denotes the number of transmissions required to recover a block successfully in any ARQ scheme, then the expected value of V is

$$
\begin{aligned}
E[V] = \quad & 1 \cdot Pr[A_o^c + A_o^e] + 2 \cdot Pr[A_o^d(E_1^c + E_1^e)] + \cdots \\
& + (i+1) \cdot Pr[A_o^d E_1^d \cdots E_{i-1}^d(E_i^c + E_i^e)] + \cdots
\end{aligned}
\tag{6.11}
$$

and the throughput efficiency of the ARQ system is given by

$$
\eta = \frac{1}{E[V]}(\frac{k}{n}).
\tag{6.12}
$$

The inequalities, $Pr[E_i^c] >> Pr[E_i^e]$ and $Pr[A_o^c] >> Pr[A_o^e]$, may, under appropriate conditions be used to obtain an excellent approximation to $E[V]$ of (6.11). The quantity k/n is the rate of the code C_o used for error detection. The expression for E[V] given in (6.11) is a general expression for the average number of transmissions in an ARQ system. However, it involves the probability of joint events which are difficult to calculate. We, therefore, adopt the following two approaches for further analysis.

Approach 1. Wang et al [6.13] obtained the lowerbound on the throughput of a type-II hybrid ARQ system by analyzing the throughput of an ARQ system which performs error correction at every *odd* retransmission and only error detection of every *even* retransmission. These expressions can be appropriately modified when an (8, 4, 3) KM code is used for error correction and are given in Appendix C. Note that this code corresponds to the simplest form of a GH-ARQ system having $m = 2$. Figures 6.7-6.9 show the throughput of such a system for $n = 500$, 1000, and 2000, respectively. These figures also show the throughput of the type-II hybrid ARQ system for the error correcting capability of the code C_1, $t_1 = 5$, 10, and 20 [6.13]. It can be seen that the throughput of the GH-ARQ system decreases slowly as it approaches 0.5. Although an exact analysis appears to be complex, it can be argued that since the receiver performs error detection

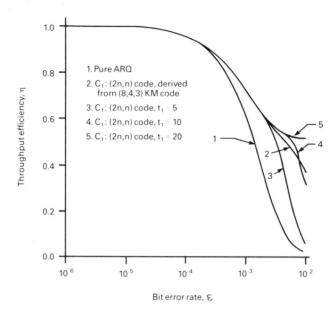

Figure 6.7: Throughput Efficiency of Selective-Repeat GH-ARQ Schemes Using Depth 2 Codes for Error Correction ($n = 500$).

as well as error correction upon the first retransmission, the probability of further retransmissions is significantly reduced and, therefore, a higher throughput is achieved. Also, as long as the probability of error correction upon the first retransmission is high, the throughput will remain close to 0.5.

Approach 2. Let us consider a GH-ARQ system, denoted by S_m, that uses an (mn, n) code C_1 for error correction. Also define a series of GH-ARQ systems S_j as the GH-ARQ systems that use the subcodes $C_1^{(j)}$ of C_1 for error-correction, $j = 1, 2, \ldots, m$. We further assume that $d_\alpha < d_\beta$, if $\alpha < \beta$, where d_θ denotes the minimum distance of the subcode $C_1^{(\theta)}$. If $E_j[V]$ is the expected number of transmissions for the GH-ARQ system S_j defined above, then we have

$$E_j[V] = Pr_j[A_o^c + A_o^e] + \sum_{i=1}^{\infty}(i+1) \cdot Pr_j[F_i], \qquad j = 1, 2, \ldots, m, \quad (6.13)$$

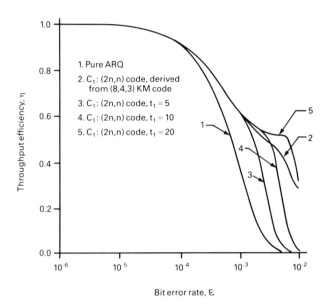

Figure 6.8: Throughput Efficiency of Selective-Repeat GH-ARQ Schemes Using Depth 2 Codes for Error Correction ($n = 1000$).

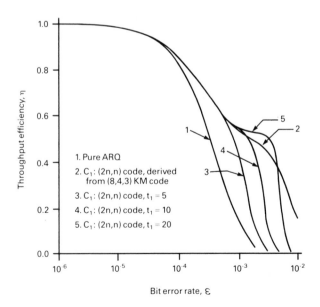

Figure 6.9: Throughput Efficiency of Selective-Repeat GH-ARQ Schemes Using Depth 2 Codes for Error Correction ($n = 2000$).

where $F_i = A_o^d E_1^d E_2^d \cdots E_{i-1}^d (E_i^c + E_i^e)$. Also, $P_{S_j}[i] = Pr_j[A_o^c + A_o^e] + \sum_{\ell=1}^{i-1} Pr_j[F_\ell]$ is the probability of the event that for GH-ARQ system S_j, the receiver recovers the data block successfully in i transmissions. Intuitively, we see that

$$P_{S_\alpha}[i] = P_{S_\beta}[i], \quad \text{for} \quad i \leq \alpha \quad \text{and} \quad i \leq \beta, \qquad (6.14)$$

and

$$P_{S_\alpha}[i] > P_{S_\beta}[i], \quad \text{for} \quad \alpha > \beta \quad \text{and} \quad i > \beta. \qquad (6.15)$$

The above equations (6.14) and (6.15) are mathematical analogs of the following statement: "The probability of the event that a block is successfully accepted in i transmissions is equal for two GH-ARQ schemes that use codes of depth α and β for error correction if $i \leq \alpha$ and $i \leq \beta$, since the two schemes behave in identically the same manner for such a value of i. Conversely, the probability of the event that a block is successfully accepted in i transmissions is higher for a GH-ARQ scheme that uses a greater depth code for error correction."

The throughput of a GH-ARQ system using a depth α code for error correction is, therefore, always higher than a GH-ARQ system that uses a depth β code if $\alpha > \beta$. The relative improvement in the throughput, however, depends on the value of $P_{S_j}[i]$. For example, if the error-rate of the channel $\varepsilon \to 0$, then, $P_{S_j}[1] \simeq Pr_j[A_o^c] = (1 - \varepsilon)^n \to 1$, and $E_j[V] \to 1$, for $j = 1, 2, \ldots, m$; therefore, all the GH-ARQ schemes perform equally well for low error rates. As ε increases, we require a larger value of i, i.e., more transmissions for a data block in order to make $P_{S_j}[i]$ close to unity, thereby increasing $E_j[V]$. Also, a GH-ARQ system $S_{\alpha+1}$ will perform significantly better than a GH-ARQ system S_α only if $P_{S_{\alpha+1}}[\alpha + 1] \gg P_{S_\alpha}[\alpha + 1]$, since $P_{S_{\alpha+1}}[\alpha] = P_{S_\alpha}[\alpha]$. Furthermore, for any GH-ARQ system S_j, if $P_{S_j}[i] \to 1$ for the smallest value of i, then the throughput of such a system will fall to at most $1/i$. For example, for the type-II hybrid ARQ system (S_2), the throughput falls to 0.5 when retransmissions take place and ε is such that $P_{S_2}[2] \to 1$. For a given channel, the depth m of the GH-ARQ scheme S_m should be selected such that $P_{S_m}[m] \to 1$ for the largest possible value of ε.

In Appendix C, we analyze the throughput of a GH-ARQ scheme using a depth 3 code C_1 by defining two systems *inferior* to the proposed system. The first inferior system A performs error correction at every *odd* retransmission using the $C_1^{(2)}$ code and only error detection at every *even* retransmission. The second inferior system B performs error correction at every *third* transmission using the C_1 code and only error detection for *other* transmissions. Since the expressions derived are similar to the expressions derived by Wang et al [6.13], only the pertinent details are given. The throughput of the actual GH-ARQ scheme is underbounded by the maximum of the throughputs of the two inferior systems.

The throughput expressions for GH-ARQ schemes employing different depth codes for error correction can also be derived in a fashion similar to the approach taken in Appendix C. Figure 6.10 shows the throughput of a GH-ARQ scheme using the (12,4,6) KM code for error correction for various values of data block length n. Such a scheme has a depth of 3, and Figure 6.10 illustrates that the throughput is close to $1/3$ even for very high values of error rate ($\varepsilon \approx 10^{-2}$) and large block-length ($n = 2000$). Similarly, Figure 6.11 shows the throughput of a GH-ARQ scheme using the (15,5,5) KM code for error correction for various values of data block length n. In the next section, we improve further on the GH-ARQ hard-decision approach taken this far.

6.8 Soft-Decision Decoding

ARQ methods predominantly use hard-decision decoding, although [6.12] briefly describes soft-decision decoding for the $(2n, n)$ case. Another soft-decision ARQ method is discussed in [6.15, 6.16]; this approach is hybrid, with each retransmission identical to the first transmission. It is well known that the use of soft decisions provides the communications engineer with an additional degree of design freedom and results, for the additive white Gaussian noise (AWGN) channel, in about a 2dB improvement in signal-to-noise ratio, E_b/N_o. When the codeword length n is small, it is possible to carry out *maximum likelihood decoding* exactly, a procedure which would prove impractical for long codes. In this section, we select a message length and a specific short KM code and illustrate by simulation the performance improvement possible for the scheme described in [6.7, 6.8] when maximum likelihood sequence estimation is employed. The general approach and philosophy of soft-decision GH-ARQ is similar to the code combining

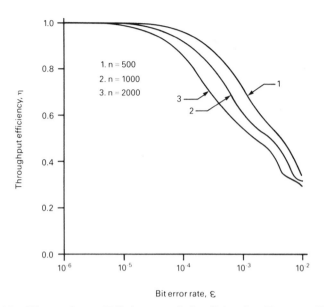

Figure 6.10: Throughput Efficiency of the Selective-Repeat GH-ARQ Schemes Using the (12,4,6) KM Code for Error Correction.

technique of Chase [6.17] and, therefore, may also be expected to work well under channel conditions far less benign than those assumed here. Additional details on the approach taken may be found in [6.9, 6.18, and 6.21].

6.8.1   Background

In this section, a brief description is provided of several elements of the soft-decision GH-ARQ scheme, in particular, the linear block codes and soft-decision method employed. A flow diagram of the receiver operation is shown in Figure 6.12; the parameter BL denotes the message block transmitted corresponding to a given set of information bits. The decision box labeled "Errors?" refers to use of the error detection code C_o; in this section, C_o is the (500,480,5) code obtained by shortening the (1023,1003,5) BCH code having generator polynomial $1 + u + u^2 + u^4 + u^5 + u^6 + u^{11} + u^{12} + u^{20}$. The message (block) length chosen is, therefore, 500 bits.

Error Correction Codes. The (n, k, d) error correction code C_1 em-

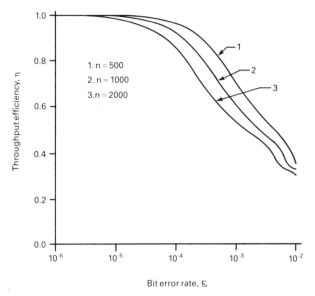

Figure 6.11: Throughput Efficiency of the Selective-Repeat GH-ARQ Schemes Using the (15,5,5) KM Code for Error Correction.

ployed is once again the (12,4,6) depth 3 KM code. The columns of the generator matrix can be partitioned as shown in (6.6); therefore, we have $M = [M_1 \mid M_2 \mid M_3]$, where $[M_1 \mid M_2]$ is the generator matrix for the (8,4,3) KM code. All submatrices M_i, $i = 1, 2, 3$, are invertible. The message is divided into contiguous groups of k-bits each; let the column vector $\mathbf{d}_\ell$ denote the ℓth group and define the subblock $\mathbf{b}_\ell^{(i)} = M_i^T \mathbf{d}_\ell$. The ith encoded block is given by

$$\mathbf{B}^{(i)} = \left[\mathbf{b}_1^{(i)T}, \mathbf{b}_2^{(i)T}, \dots, \mathbf{b}_\ell^{(i)T}, \dots, \mathbf{b}_L^{(i)T} \right]^T, \qquad (6.16)$$

where $L = N/n (= 125)$. The encoder first uses M_1 to generate $\mathbf{B}^{(1)}$. If a NACK is returned, the encoder uses M_2 to generate $\mathbf{B}^{(2)}$ and, finally, for this depth 3 scheme, if a NACK is still returned, the encoder uses M_3 to generate $\mathbf{B}^{(3)}$. Since the M_i are invertible, $\mathbf{d}_\ell = \left(M_i^T \right)^{-1} \mathbf{b}_\ell^{(i)}$ and any $\mathbf{d}_\ell$ can be recovered from any block. Encoding is implemented as described in Section 5.2.3 using linear feedback shift registers (LFSR) and mod-2 adders in a straightforward manner.

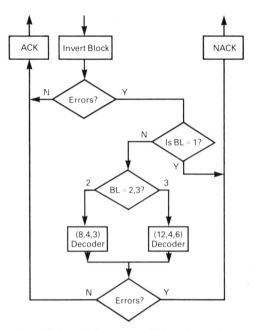

Figure 6.12: Flow Diagram of Receiver Operation.

Soft-Decision Decoding. The ith received block is denoted by $\tilde{\mathbf{B}}^{(i)}$ and is partitioned into L contiguous subblocks. As shown in Figure 6.12, each subblock is inverted and subject to error detection. When errors are detected in $\tilde{\mathbf{B}}^{(1)}$, the second block is requested. If the erroneous block is other than the first, it is combined with the previous block(s) for soft-decision decoding. The combining operation is shown in Figure 6.13 for the ℓth received subblock. The elements of any $\tilde{\mathbf{B}}^{(i)}$ are taken as the analog demodulator output values, which we simply denote as y_j. Maximum likelihood sequence estimation is employed for decoding [6.19]. For subblocks combined as above, the decoder forms the K decision variables

$$U_i = \sum_{j=1}^{n}(2c_{ij} - 1)y_j, \qquad i = 1, 2, \ldots, K, \qquad (6.17)$$

where $n = 8$ or 12, depending on the depth; c_{ij} denotes the bit in the jth position of the ith codeword; and K is the number of codewords. We assume *binary phase-shift-keyed* (BPSK) signaling; thus,

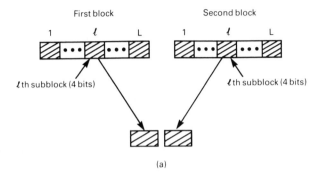

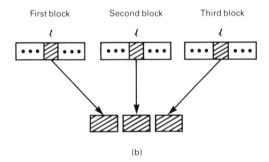

Figure 6.13: Combining Method for Soft-Decision Decoding: (a) ℓth Input to (8,4,3) Decoder, (b) ℓth Input to (12,4,6) Decoder.

the y_j, $j = 1, 2, \ldots, n$, represent the n sampled outputs for any particular codeword of a filter matched to the binary PSK waveform. For the AWGN channel, $y_j = \pm 2\alpha E_b + n_j$, where α is the channel attenuation; E_b is the signal energy per bit; and n_j is normally distributed as $\mathcal{N}(0, 2E_bN_o)$, with N_o the equivalent low-pass noise power spectral density.

The largest (over i) of the U_i's specifies the desired codeword. The above decoding method is employed only when inversion has failed to produce error-free decoding. Since $K = 2^k(= 16)$, it is possible to perform the computations required by (6.17) (even at high bit rates, if a requisite number of subblocks are decoded in parallel) without resort to approximation.

6.8.2   System Performance

In the previous sections, it was shown that the hard-decision GH-ARQ scheme offers significant performance improvement over hard-decision type-II hybrid ARQ methods. The purpose of this section is to provide analytical and simulation results on soft-decision GH-ARQ reliability and throughput efficiency for the specific example described in Section 6.8.1. Throughput efficiency simulation results were based on averages over at least 10^4 blocks, each containing 500 randomly selected bits.

Error Probabilities. The previous results indicate that for the codes $\mathcal{C}_o$ and $\mathcal{C}_1$ as selected here, the probability of undetected error for the first transmission and first and second retransmissions will be less than 10^{-6} for a channel bit-error-rate (BER) less than 0.5. The channel BER, $\varepsilon = (1/2)ERFC(\sqrt{\gamma_b})$, where $\gamma_b = \alpha^2 E_b/N_o$. In order to analytically compute the throughput efficiency, it is important to evaluate the post-decoding bit error probability P_b for the (12,4,6) and (8,4,3) codes. It is not difficult to show that

$$P_b = [2n(K-1)]^{-1} \sum_{w=d}^{n} ERFC(\sqrt{\gamma_b R w})n(\omega), \qquad (6.18)$$

where $R = k/n$ and $n(\omega)$ is the number of codewords having Hamming weight equal to ω. The weight distributions for both codes were computed, with the post-decoding bit error probabilities shown in Figure 6.14.

Throughput Efficiency. Throughput efficiency η is given by (6.12) and defined as the rate of the code C_o (0.96) divided by the expected number of transmissions required to recover a block successfully. Calculation of throughput efficiency may be carried out by reference to Figure 6.15, where P_1 is the probability that inversion contains detectable errors and P_i is the probability that depth-i decoding contains detectable errors, $i = 2, 3$. The probabilities P_i, $i = 2, 3$, may be found using the channel BER, ε; the number of errors detected by C_o; and (6.18). Let Q_i denote the probability of successfully recovering a block on the ith transmission; this results in the following expression for the throughput efficiency,

$$\eta = 0.96(1 - P_1^3 P_2 P_3)^2 / [Q_1(1 + 2P_1^3 P_2 P_3) + Q_2(2 + P_1^3 P_2 P_3) + 3Q_3].$$
(6.19)

When the channel is quiet $\eta \cong (1 + 2P_1)^{-1}$, with P_1 depending only on the channel BER and the code C_o. Table 6.1 is a compilation of the soft-decision GH-ARQ throughput efficiency versus channel BER, obtained both from (6.19) and by simulation. The simulation and analytical results are seen to be in close agreement.

6.8.3 Performance Comparisons

Two comparisons of throughput efficiency are made: (1) against the hard-decision GH-ARQ method described earlier and (2) against the soft-decision ARQ scheme found in [6.15, 6.16]. The probability of undetected error is maintained at 10^{-6} for these simulation studies. The importance of fixing one parameter while another is varied for system performance comparison is pointed out in [6.20]. Figure 6.16 illustrates the first comparison and dramatically shows the improvement possible for poor channel conditions ($\varepsilon > 10^{-4}$) when soft-decision GH-ARQ is used with a depth of 3. Before presenting the next comparison, a brief description of the scheme used in [6.15, 6.16] is in order. This method uses $2n_L$ quantization levels, where $n_L = 2$. Denote by $y_{m,j}$ the quantized demodulator output in the jth symbol interval for the mth transmission of a codeword. The decoder forms a vector $\mathbf{z}_m$ whose jth component is

$$z_{m,j} = z_{m-1,j} + y_{m,j}, \quad j = 1, 2, \ldots, n,$$
(6.20)

with $z_{o,j} = 0$. After each set of transmissions (three, in this case), the

Table 6.1: Throughput Efficiency for Soft-Decision GH-ARQ.

Channel BER	η(anal.)	η(simul.)
1.00×10^{-6}	1.00	1.00
1.00×10^{-5}	1.00	1.00
1.00×10^{-4}	1.00	1.00
1.00×10^{-3}	.99	1.00
2.51×10^{-3}	.98	1.00
3.98×10^{-3}	.83	.77
1.00×10^{-2}	.69	.74
3.16×10^{-2}	.42	.38
4.64×10^{-2}	.30	.27
6.81×10^{-2}	.11	.14
1.00×10^{-1}	.05	.06
2.15×10^{-1}	.02	.01

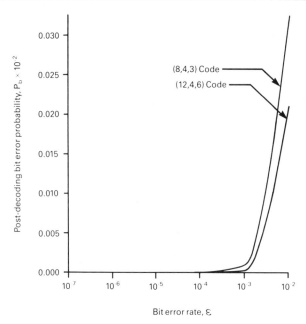

Figure 6.14: Post-Decoding Bit Error Probabilities for the (8,4,3) and (12,4,6) Codes.

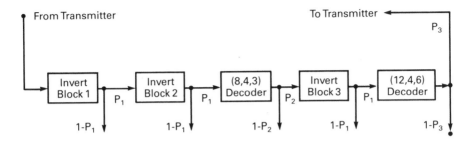

Figure 6.15: Decoding Model Used for the Performance Analysis

decoder performs a hard decision on z_m and examines the resulting binary vector to see if it is a codeword. If it is not, a new set of transmissions for the same codeword is required. The comparison with this method is presented in Figure 6.17. To be consistent with [6.16], the abcissa is P_B, the channel *block* error probability. This probability is defined as $P_B = 1 - (1 - \varepsilon)^n$. The superior performance of soft-decision GH-ARQ is due to two reasons: (1) a greater number of quantization levels is employed (although beyond eight levels, the improvement is slight) and (2) the power of the error correction increases with retransmission.

6.8.4 Soft-Decision GH-ARQ Code Alternatives

Since we have concentrated in the above study on the (12,4,6) KM code and its use in a soft-decision GH-ARQ scheme, the question naturally arises as to the performance of other KM codes, of depth 3 and even depth 4. Figure 6.18 provides a comparison of the throughput efficiencies for a number of depth 3 KM codes, and Figure 6.19 shows the corresponding post-decoding bit-error probabilities. It is clear that the (21,7,6) code offers the best throughput efficiency of the codes examined; in fact, the throughput efficiency is 98% or higher for an extremely wide range of channel BER.

A number of depth 4 systems were also compared. This category includes the (24,6,9), (28,7,10), and (32,8,9) KM codes. The (28,7,10) KM code is obtained from the (27,7,9) KM code by adding an overall parity-check bit. As shown in Figure 6.20, of this group, the (28,7,10) code offered the best throughput efficiency. Figure 6.21 shows the corresponding post-decoding bit-error probabilities. Little performance

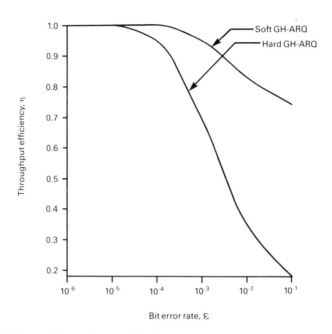

Figure 6.16:   Throughput Efficiency Comparison of Hard- and Soft-Decision GH-ARQ Schemes.

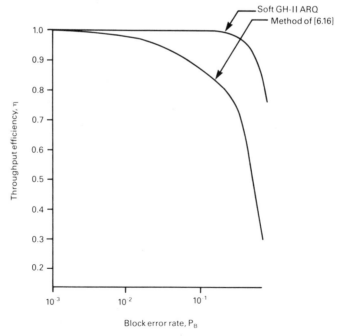

Figure 6.17:   Throughput Efficiency Comparison of Soft-Decision GH-ARQ and the Method of [6.16].

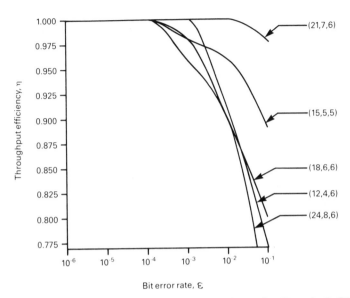

Figure 6.18: Throughput Efficiency Comparison for Depth 3 GH-ARQ Systems.

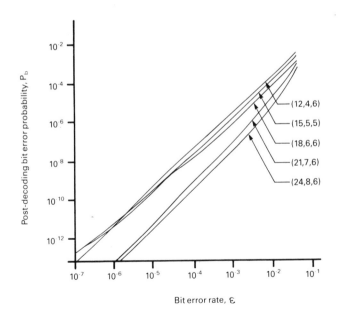

Figure 6.19: Post-Decoding Bit Error Probability for Depth 3 GH-ARQ Systems.

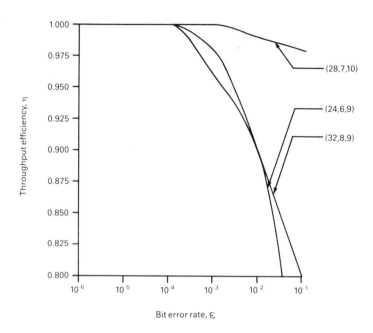

Figure 6.20: Throughput Efficiency Comparison for Depth 4 GH-ARQ Systems.

improvement of depth 4 systems is observed over depth 3 systems; this is due primarily to the fact that an AWGN channel was assumed, as described in Section 6.8.1.

Experimental work is presently underway using the (28,7,10) KM code. Since this code contains the (21,7,6) code, a depth 3 system is easily obtained from the depth 4 system by simply ignoring the last block of the generator matrix. Preliminary results indicate that these depth 3 and depth 4 systems significantly improve throughput efficiency and reliability of line-of-sight (LOS) microwave links under moderately heavy fading conditions. The links used form elements of a larger telecommunications network.

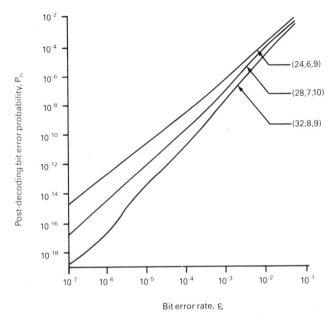

Figure 6.21: Post-Decoding Bit Error Probability for Depth 4 GH-ARQ Systems.

6.9 References

[6.1] W. W. Peterson and E. J. Weldon, Jr., *Error Correcting Codes*, The MIT Press, 1978.

[6.2] S. Lin, D. J. Costello, Jr. and M. J. Miller, "Automatic-Repeat-Request Error-Control Schemes," *IEEE Communications Magazine*, vol. 22, no. 12, pp. 5-17, Dec. 1984.

[6.3] S. Lin and P. S. Yu, "A Hybrid ARQ Scheme with Parity Retransmission for Error Control of Satellite Channels," *IEEE Trans. Commun.*, vol. COM-30, pp. 1701-1719, July 1982.

[6.4] A. R. K. Sastry, "Improving Automatic Repeat-Request (ARQ) Performance on Satellite Channels under High Error Rate Conditions," *IEEE Trans. Commun.*, vol. COM-23, pp. 436-439, April 1975.

[6.5] S. Lin and D. J. Costello, Jr., *Error Control Coding: Fundamentals and Applications*, Prentice Hall, 1983.

[**6.6**] E. Y. Rocher and R. L. Pickholtz, "An Analysis of the Effectiveness of Hybrid Transmission Schemes," *IBM Journal of Research and Development*, pp. 426-433, July 1970.

[**6.7**] H. Krishna and S. D. Morgera, "A New Error Control Scheme for Hybrid ARQ Systems," in Proc. *IEEE Intern. Conf. on Commun.*, Toronto, Canada, pp. 637-645, June 22-25, 1986.

[**6.8**] H. Krishna and S. D. Morgera, "A New Error Control Scheme for Hybrid ARQ Systems," *IEEE Trans. Commun.*, vol. COM-35, pp. 981-990, Oct. 1987.

[**6.9**] S.D. Morgera and V. K. Oduol, "Soft-Decision Decoding Applied to the Generalized Type-II Hybrid ARQ Scheme," in Proc. *IEEE Intern. Conf. on Commun.*, Philadelphia, PA, June 12-15, 1988.

[**6.10**] P. S. Yu and S. Lin, "An Efficient Selective-Repeat ARQ Scheme for Satellite Channels and its Throughput Analysis," *IEEE Trans. Commun.*, vol. COM-29, pp. 353-363, March 1981.

[**6.11**] D. M. Mandelbaum, "Adaptive-Feedback Coding Scheme Using Incremental Redundancy," *IEEE Trans. Inform. Theory*, vol. IT-20, pp. 388-389, May 1974.

[**6.12**] J.J. Metzner, "Improvements in Block-Retransmission Schemes," *IEEE Trans. Commun.*, vol. COM-27, pp. 525-532, Feb. 1979.

[**6.13**] Y. M. Wang and S. Lin, "A Modified Selective-Repeat Type-II Hybrid ARQ System and its Performance Analysis," *IEEE Trans. Commun.*, vol. COM-31, pp. 593-607, May 1983.

[**6.14**] T. Kasami, T. Klove and S. Lin, "Linear Block Codes for Error Detection," *IEEE Trans. Inform. Theory*, vol. IT-29, pp. 131-136, Jan. 1983.

[**6.15**] G. Benneli, "Efficient ARQ Technique Using Soft Demodulation," *Electronics Letters*, vol. 22, no. 4, pp. 205-206, Feb. 1986.

[**6.16**] G. Benneli, "An ARQ Scheme with Memory and Soft-Error Detectors," *IEEE Trans. Commun.*, vol. COM-33, pp. 285-288, March 1985.

[6.17] D. Chase, "Code Combining - a Maximum-Likelihood Approach for Combining an Arbitrary Number of Noisy Packets," *IEEE Trans. Commun.*, vol. COM-33, pp. 385-393, May 1985.

[6.18] V. K. Oduol, "A Generalised Type-II Hybrid ARQ Scheme with Soft-Decision Decoding," M.Eng. Thesis, McGill University, Dept. of Elect. Engin., Montreal, March 1987.

[6.19] J. G. Proakis, *Digital Communications*, McGraw-Hill, 1983.

[6.20] A. R. K. Sastry and L. N. Kanal, "Hybrid Error Control Using Retransmission and Generalized Burst-Trapping Codes," *IEEE Trans. Commun.*, vol. COM-24, pp. 385-393, April 1976.

[6.21] S.D. Morgera and V.K. Oduol, "Soft-Decision Decoding Applied to the Generalized Type II Hybrid ARQ Scheme," *IEEE Trans. Commun.* (to appear).

6.10   Appendix A

A Menu of Error-Correcting Codes for GH-ARQ Systems

In this appendix, we list several KM codes that can be used for adaptive error control in a GH-ARQ scheme. Also described are some of the properties that can be used to identify and generate these codes from the basic design procedure. Note that the generator matrices are completely characterized by the dimension k, design distance d, the choice of modulo polynomial $P(u)$, and the wraparound s.

$(15,5,5)$ KM code. Here $P(u) = u^3(u^2 + 1)(u^2 + u + 1)$; $s = 2$. The generator matrix M is given by

$$M = \left[\begin{array}{ccccc|ccccc|ccccc} 1 & 1 & 0 & 1 & 0 & 0 & 1 & 1 & 0 & 1 & 0 & 0 & 1 & 0 & 0 \\ 0 & 0 & 1 & 0 & 0 & 1 & 1 & 1 & 1 & 1 & 0 & 1 & 0 & 0 & 0 \\ 0 & 1 & 0 & 1 & 0 & 0 & 0 & 1 & 1 & 0 & 1 & 1 & 1 & 0 & 0 \\ 0 & 0 & 1 & 1 & 0 & 0 & 0 & 1 & 0 & 1 & 0 & 0 & 0 & 1 & 1 \\ 0 & 1 & 0 & 0 & 1 & 0 & 0 & 1 & 1 & 1 & 0 & 0 & 0 & 0 & 1 \end{array}\right] \quad (6.21)$$

$$= [M_1 \mid M_2 \mid M_3].$$

Features:

a. The matrices M_1, M_2, and M_3 are invertible.

b. $[M_1|M_2]$ forms the generator matrix of a $(10,5,3)$ KM code corresponding to $P(u) = u^2(u^2 + 1)(u^2 + u + 1)$; $s = 1$.

No direct comparison of the above described code with the $(15,5,7)$ BCH code [6.1] appears possible since it is not known if the $(15,5,7)$ BCH code satisfies the requirements to be useful for the application under consideration.

$(18,6,6)$ KM Code. Here $P(u) = u^2(u^2 + 1)(u^2 + u + 1)(u^3 + u^2 + 1)$; $s = 2$. The generator matrix M is given by

$$M = \begin{bmatrix} 1 & 1 & 0 & 0 & 0 & 1 & 1 & 0 & 1 & 1 & 0 & 1 & 0 & 1 & 1 & 0 & 0 & 0 \\ 0 & 0 & 1 & 1 & 0 & 0 & 1 & 0 & 1 & 0 & 1 & 1 & 1 & 1 & 0 & 1 & 0 & 0 \\ 0 & 0 & 0 & 1 & 0 & 1 & 1 & 1 & 0 & 1 & 1 & 0 & 0 & 0 & 1 & 0 & 0 & 0 \\ 0 & 1 & 0 & 1 & 0 & 1 & 1 & 1 & 1 & 0 & 0 & 1 & 0 & 0 & 0 & 1 & 0 & 0 \\ 0 & 1 & 1 & 0 & 0 & 0 & 1 & 1 & 0 & 0 & 1 & 1 & 0 & 0 & 1 & 0 & 1 & 1 \\ 0 & 1 & 1 & 1 & 1 & 1 & 1 & 0 & 0 & 1 & 1 & 0 & 0 & 0 & 0 & 1 & 0 & 1 \end{bmatrix}$$

$$(6.22)$$

$$= [M_1 \mid M_2 \mid M_3].$$

Features:

a. The matrices M_1, M_2, and M_3 are invertible.

b. $[M_1|M_2]$ forms the generator matrix of a (12,6,3) KM code corresponding to $P(u) = u(u+1)(u^2+u+1)(u^3+u^2+1)$; $s = 1$.

(24,6,9) KM Code. If we append one more block, M_4, to the generator matrix of the (18,6,6) KM code of (6.22), we obtain the (24,6,9) KM code. Such a block corresponds to the computation for the factor polynomial (u^3+u+1) and is given by

$$M_4 = \begin{bmatrix} 1 & 0 & 0 & 1 & 0 & 1 \\ 0 & 1 & 0 & 1 & 1 & 0 \\ 0 & 0 & 1 & 0 & 1 & 1 \\ 1 & 1 & 0 & 0 & 1 & 1 \\ 0 & 1 & 1 & 1 & 0 & 1 \\ 1 & 1 & 1 & 0 & 0 & 0 \end{bmatrix}$$

Note that M_4 is not invertible.

(28,7,10) Extended KM Code. Such a code is obtained by adding an overall parity-check to the (27,7,9) KM code having $P(u) = u^2(u^2+1)(u^2+u+1)(u^3+u^2+1)(u^3+u+1)$; $s = 3$. The generator matrix M is given by

$$M = [M_1|M_2|M_3|M_4],$$

where

$$M_1 = \begin{bmatrix} 1 & 1 & 1 & 1 & 0 & 0 & 0 \\ 0 & 1 & 0 & 0 & 1 & 1 & 0 \\ 0 & 1 & 1 & 0 & 0 & 1 & 0 \\ 0 & 1 & 1 & 1 & 0 & 1 & 0 \\ 0 & 1 & 0 & 1 & 1 & 0 & 0 \\ 0 & 1 & 1 & 1 & 1 & 1 & 0 \\ 0 & 1 & 1 & 0 & 1 & 0 & 1 \end{bmatrix},$$

$$M_2 = \begin{bmatrix} 0 & 1 & 0 & 1 & 1 & 0 & 0 \\ 1 & 1 & 0 & 1 & 0 & 0 & 0 \\ 1 & 0 & 1 & 0 & 1 & 0 & 0 \\ 0 & 1 & 1 & 1 & 0 & 0 & 0 \\ 1 & 1 & 1 & 0 & 0 & 0 & 0 \\ 1 & 0 & 0 & 0 & 1 & 1 & 1 \\ 0 & 1 & 1 & 1 & 1 & 0 & 1 \end{bmatrix},$$

$$M_3 = \begin{bmatrix} 0 & 1 & 1 & 0 & 0 & 0 & 0 \\ 1 & 1 & 0 & 1 & 0 & 0 & 0 \\ 0 & 0 & 1 & 0 & 0 & 0 & 0 \\ 0 & 0 & 0 & 1 & 0 & 0 & 0 \\ 0 & 0 & 1 & 0 & 1 & 1 & 1 \\ 0 & 0 & 0 & 1 & 0 & 0 & 1 \\ 0 & 0 & 1 & 0 & 0 & 1 & 0 \end{bmatrix},$$

$$M_4 = \begin{bmatrix} 1 & 0 & 0 & 1 & 0 & 1 & 0 \\ 0 & 1 & 0 & 1 & 1 & 0 & 0 \\ 0 & 0 & 1 & 0 & 1 & 1 & 0 \\ 1 & 1 & 0 & 0 & 1 & 1 & 0 \\ 0 & 1 & 1 & 1 & 0 & 1 & 0 \\ 1 & 1 & 1 & 0 & 0 & 0 & 0 \\ 1 & 0 & 1 & 1 & 1 & 0 & 1 \end{bmatrix}. \tag{6.23}$$

Features:

a. The matrices M_1, M_2, and M_3 are invertible.

b. $[M_1|M_2]$ forms the generator matrix of a (14,7,3) KM code; $P(u) = u(u+1)(u^2+u+1)(u^3+u^2+1)$; $s = 2$.

c. $[M_1|M_2|M_3]$ forms the generator matrix of a $(21,7,6)$ KM code;
$P(u) = u^2(u^2+1)(u^2+u+1)(u^3+u^2+1);\ s = 3.$

Note: The decoder configuration of the extended KM code is the same as that of the associated KM code followed by a simple parity-check.

(32,8,9) KM Code. In this case, $P(u) = u^3(u^2+1)(u^2+u+1)(u^3+u^2+1)(u^3+u+1);\ s = 3.$ The generator matrix M is given by

$$M = [M_2|M_2|M_3|M_4],$$

where

$$M_1 = \begin{bmatrix} 1 & 1 & 0 & 0 & 1 & 0 & 0 & 0 \\ 0 & 0 & 1 & 1 & 0 & 1 & 1 & 0 \\ 0 & 1 & 0 & 1 & 0 & 0 & 1 & 0 \\ 0 & 0 & 1 & 0 & 1 & 0 & 1 & 0 \\ 0 & 1 & 0 & 1 & 1 & 1 & 0 & 0 \\ 0 & 0 & 1 & 1 & 1 & 1 & 1 & 0 \\ 0 & 1 & 0 & 0 & 0 & 1 & 0 & 0 \\ 0 & 0 & 1 & 1 & 1 & 0 & 0 & 1 \end{bmatrix},$$

$$M_2 = \begin{bmatrix} 1 & 1 & 1 & 0 & 1 & 1 & 0 & 0 \\ 1 & 0 & 1 & 0 & 1 & 0 & 0 & 0 \\ 1 & 1 & 0 & 1 & 0 & 1 & 0 & 0 \\ 1 & 1 & 1 & 1 & 1 & 0 & 0 & 0 \\ 1 & 0 & 1 & 1 & 0 & 0 & 0 & 0 \\ 1 & 1 & 0 & 0 & 0 & 1 & 0 & 0 \\ 1 & 1 & 1 & 1 & 1 & 1 & 1 & 1 \\ 1 & 0 & 1 & 0 & 1 & 1 & 0 & 1 \end{bmatrix},$$

$$M_3 = \begin{bmatrix} 0 & 1 & 0 & 0 & 0 & 1 & 0 & 0 \\ 1 & 1 & 0 & 0 & 0 & 0 & 1 & 1 \\ 0 & 0 & 0 & 0 & 0 & 0 & 0 & 1 \\ 0 & 0 & 0 & 0 & 0 & 1 & 1 & 1 \\ 0 & 0 & 0 & 0 & 0 & 0 & 1 & 0 \\ 0 & 0 & 1 & 1 & 1 & 1 & 1 & 0 \\ 0 & 0 & 0 & 1 & 0 & 1 & 0 & 1 \\ 0 & 0 & 0 & 0 & 1 & 1 & 0 & 0 \end{bmatrix},$$

$$M_4 = \begin{bmatrix} 0 & 0 & 1 & 1 & 0 & 1 & 1 & 0 \\ 0 & 1 & 0 & 1 & 0 & 0 & 0 & 1 \\ 1 & 1 & 1 & 0 & 1 & 1 & 0 & 0 \\ 0 & 0 & 0 & 0 & 0 & 1 & 0 & 0 \\ 0 & 0 & 0 & 1 & 1 & 1 & 1 & 1 \\ 0 & 0 & 0 & 0 & 1 & 0 & 0 & 1 \\ 0 & 0 & 0 & 1 & 1 & 0 & 0 & 0 \\ 0 & 0 & 0 & 1 & 0 & 1 & 1 & 1 \end{bmatrix} \tag{6.24}$$

Features:

a. The matrices M_1, M_2, M_3, and M_4 are invertible.

b. $[M_1|M_2]$ forms the generator matrix of a (16,8,3) KM code; $P(u) = u(u^2 + 1)(u^2 + u + 1)(u^3 + u^2 + 1)$; $s = 2$.

c. $[M_1|M_2|M_3]$ forms the generator matrix of a (24,8,6) KM code; $P(u) = u^2(u^2 + 1)(u^2 + u + 1)(u^3 + u^2 + 1)$; $s = 3$.

Note: The first 30 columns of the matrix given in (6.24) are the same as the (30,8,9) KM code. The last two columns are introduced in such a way as to render M_4 invertible and the overall length n a multiple of the dimension k.

6.11 Appendix B

Reliability of GH-ARQ Systems

In this appendix, it is shown that the GH-ARQ system provides the same order of reliability as an ARQ system using a code C_o for error detection and having probability of undetected error, P_e. In order to see this, we shall define a number of important events.

Let A_o^c, A_o^d, and A_o^e be the events that a data block contains no errors, detectable errors, and undetectable errors, respectively, in its first transmission. Let B_i^c, B_i^d, and B_i^e be the events that the ith retransmission ($i \geq 1$) for a data block contains no errors, detectable errors, and undetectable errors, respectively. Note that the ith retransmission corresponds to a total of $i + 1$ transmissions for a given data block. Finally, upon the ith retransmission, let Q_i^c, Q_i^d, and Q_i^e be the events that the data block obtained by decoding the blocks received up to the ith retranmission contains no errors, detectable errors, and undetectable errors, respectively. Upon the ith retransmissions, the events E_i^c, E_i^d, and E_i^e corresponding, respectively, to the event that the receiver recovers the data block correctly; the event that the receiver cannot recover the data block, detects the presence of errors, and requests the next retranmission; and the event that the receiver accepts an erroneous data block may easily be represented in terms of the above events and elementary set operations as

$$
\begin{aligned}
E_i^c &= B_i^c \cup B_i^d Q_i^c, \\
E_i^d &= B_i^d Q_i^d, \\
E_i^e &= B_i^e \cup B_i^d Q_i^e.
\end{aligned}
$$

If we let E denote the event that the receiver of a GH-ARQ system accepts a data block containing undetectable errors, then the probability of the event E is given by

$$
Pr[E] = Pr[A_o^e] + \sum_{i=1}^{\infty} Pr[A_o^d E_1^d \cdots E_{i-1}^d E_i^e]. \tag{6.25}
$$

The reliability of a GH-ARQ system may be characterized by $Pr[E]$. The computation of the probabilities of joint events as required in (6.25) is difficult; however, it is possible to upperbound each of the terms in (6.25). This is the approach we take.

In a GH-ARQ scheme having depth m, error detection is performed on the basis of the codes having $\tilde{H}_i$, $i = 1, 2, \ldots, m$, as the parity-check matrix. If P_{e_i} is the probability of undetected error for such codes, let

$$
\begin{aligned}
P_e &= \max\{P_{e_i} : i = 1, 2, \ldots, m\} \\
P_f &= \min\{P_{e_i} : i = 1, 2, \ldots, m\}
\end{aligned}
$$

It is clear that $Pr[A_o^e] \leq P_e$ and that $Pr[B_i^e] \leq P_e$, $i = 1, 2, \ldots$. We also note that the probability that errors are detected in any transmission is upperbounded by $P_d = 1 - P_c - P_f$.

We now consider each term, $Pr[A_o^d E_1^d \cdots E_{i-1}^d E_i^e]$, in the summation of (6.25). Each summand may be upperbounded in the following manner:

$$
\begin{aligned}
Pr[A_o^d E_1^d \cdots E_{i-1}^d E_i^e] &\leq Pr[A_o^d B_1^d \cdots B_{i-1}^d E_i^e] \\
&\leq Pr[A_o^d B_1^d \cdots B_{i-1}^d]\, Pr[E_i^e | A_o^d B_1^d \cdots B_{i-1}^d] \\
&\leq P_d^i Pr[B_i^e \cup B_i^d Q_i^e | A_o^d B_1^d \cdots B_{i-1}^d] \\
&= P_d^i \left\{ Pr[B_i^e] + Pr[B_i^d] Pr[Q_i^e | A_o^d B_1^d \cdots B_i^d] \right\} \\
&\leq P_d^i \left\{ P_e + P_d Pr[Q_i^e | A_o^d B_1^d \cdots B_i^d] \right\}, \quad (6.26)
\end{aligned}
$$

where P_d^i is the probability that a data block contains detectable errors in each of i transmissions. Since the decoded data block is checked for the presence of errors for every retransmission, we have

$$
Pr[Q_i^e | A_o^d B_1^d \cdots B_i^d] \leq P_e. \tag{6.27}
$$

Substitution of (6.27) into (6.26) results in the desired upper bound for each summand of (6.25), i.e.,

$$
Pr[A_o^d E_1^d \cdots E_{i-1}^d E_i^e] \leq P_d^i (1 + P_d) P_e \tag{6.28}
$$

Using (6.28) and the fact that $Pr[A_o^e] \leq P_e$, the upper bound on $Pr[E]$ may be obtained as

$$Pr[E] \leq P_e + \sum_{i=1}^{\infty} P_d^i(1 + P_d)P_e$$

$$= (1 + P_d^2)P_e(P_c + P_f)^{-1}. \tag{6.29}$$

For a GH-ARQ system, $P_d < 1$ and $P_c >> P_e$; therefore, $1 + P_d^2 < 2$ and $1 + (P_e - P_f)(P_c + P_f)^{-1} \cong 1$. These expressions permit a simplication of (6.29) to

$$Pr[E] \leq 2P_e(P_c + P_e)^{-1}, \tag{6.30}$$

which is the final form for the upper bound.

For a pure ARQ system using a code C_o for error detection, the probability of the event E is given by

$$Pr[E]_{ARQ} = P_e(P_c + P_e)^{-1}, \tag{6.31}$$

a result which may be found in [6.5]. Comparing (6.30) and (6.31), we see that $Pr[E] \leq 2Pr[E]_{ARQ}$ and that the GH-ARQ scheme provides the same order of reliability as a pure ARQ scheme using a code C_o for error detection and having P_e as the probability of undetected error.

6.12   Appendix C

Throughput Efficiency of GH-ARQ Systems

In this appendix, expressions which can be used to compute the throughput of a GH-ARQ scheme using a depth $m = 3$ code C_1 for adaptive error correction are presented. The generator matrix of C_1 has the form

$$G = [G_1|G_2|G_3].$$

Let C_1 be derived from a $(3\ell', \ell')$ KM code having t_1 as its error-correcting capability, using (6.7). Also, Let $C_1^{(2)}$ be the code having $[G_1|G_2]$ as its generator matrix. Clearly, $C_1^{(2)}$ is obtained from the $(2\ell', \ell')$ code having t_2 as its error-correcting capability. Note that $t_1 > t_2$.

In order to compute the throughput of the GH-ARQ system under consideration, we define two *inferior* GH-ARQ systems A and B. The first inferior system A performs error correction at every odd retransmission using $C_1^{(2)}$ and error detection at every even retransmission. The second inferior system B performs error correction at every third transmission using C_1 and only error detection for other transmissions. This approach is very similar to the approach in [6.13] and, therefore, only the pertinent details are given.

In the GH-ARQ scheme, error detection is performed on the basis of the codes having $\tilde{H}_i$, $i = 1, 2, 3$, as the parity-check matrix. If P_{e_i} is the probability of undetected error for such codes, let

$$P_e = \max\{P_{e_i} : i = 1, 2, 3\}.$$

In the subsequent analysis, it is assumed that P_e satisfies the bound given in (6.10) and, therefore, can be made arbitrarily small. If P_c is the probability that the jth transmission of a block is received error-free, then

$$P_c = (1 - \varepsilon)^n.$$

Also, the probability that errors are detected in any transmission is at least $1 - P_c - P_e$. Since the analysis for the systems A and B is somewhat similar, we enclose the description for system B within parentheses along with the description for system A.

System A (B). For the two (three) consecutively received blocks R_{i-1} and R_i (R_{i-2}, R_{i-1}, and R_i) for the data block I, let

$q_0 =$ Probability of the event that correct decoding takes place based on $C_1^{(2)}$ (C_1);

$y =$ Probability of the event that correct decoding takes place based on $C_1^{(2)}$ (C_1) and at least one of R_{i-1} and R_i(R_{i-2}, R_{i-1}, and R_i) is error-free;

$q_1 =$ Conditional probability of correct decoding based on $C_1^{(2)}$(C_1), given that R_{i-1} and R_i(R_{i-2}, R_{i-1}, and R_i) are detected in error.

The probability of correctly obtaining I from R_{i-1} and R_i (R_{i-2}, R_{i-1}, and R_i) given that R_{i-1} (R_{i-2} and R_{i-1}) is detected in error is, therefore, given by

$$P_t = P_e + (1 - P_c)q_1.$$

For system A, $E[V]$ is underbounded by

$$E[V]_A \leq \frac{2 - P_c}{P_c + P_t - P_c P_t}. \tag{6.32}$$

The probabilities q_0, y, and q_1 may be evaluated using the expressions

$$q_0 = \left[\sum_{j=0}^{t_2} \binom{2\ell'}{j} \varepsilon^j (1 - \varepsilon)^{2\ell' - j} \right]^{n/\ell'}, \tag{6.33}$$

$$y = (1 - \varepsilon)^n \left\{ 2 \left[\sum_{j=0}^{t_2} \binom{\ell'}{j} \varepsilon^j (1 - \varepsilon)^{\ell' - j} \right]^{n/\ell'} - (1 - \varepsilon)^n \right\}, \tag{6.34}$$

and

$$q_1 = \frac{1}{1 - y}(q_0 - y). \tag{6.35}$$

If $C_1^{(2)}$ is the (8,4,3) code (as is the case for the GH-ARQ scheme based on the (12,4,6) code), $\ell' = 4$ and $t_2 = 1$.

For system B, $E[V]$ is underbounded by

$$E[V]_B \le [P_c(1 + 2a_1) + a_2(2 + a_1) + 3a_3] \cdot \frac{1}{(1 - a_1)^2} \ , \qquad (6.36)$$

where
$$a_1 = (1 - P_c)^2(1 - P_t),$$
$$a_2 = (1 - P_c)P_c, \text{ and}$$
$$a_3 = (1 - P_c)^2 P_t.$$

The probabilities q_0 and y are evaluated as follows:

$$q_0 = \left[\sum_{j=0}^{t_1} \binom{3\ell'}{j} \varepsilon^j (1 - \varepsilon)^{3\ell'-j} \right]^{n/\ell'} \qquad (6.37)$$

and

$$y = (1 - \varepsilon)^n \left[3 \left\{ \sum_{j=0}^{t_1} \binom{2\ell'}{j} \varepsilon^j (1 - \varepsilon)^{2\ell'-j} \right\}^{n/\ell'} \right.$$

$$+ 3(1 - \varepsilon)^n \left\{ \sum_{j=0}^{t_1} \binom{\ell'}{j} \varepsilon^j (1 - \varepsilon)^{\ell'-j} \right\}^{n/\ell'}$$

$$\left. - 5(1 - \varepsilon)^{2n} \right]. \qquad (6.38)$$

The expression for q_1 is the same as (6.35). If C_1 is the (12,4,6) code, $\ell' = 4$ and $t_1 = 2$.

Index